Structural Identification and Damage Detection using Genetic Algorithms

Structures and Infrastructures Series

ISSN 1747-7735

Book Series Editor:

Dan M. Frangopol

Professor of Civil Engineering and
Fazlur R. Khan Endowed Chair of Structural Engineering and Architecture
Department of Civil and Environmental Engineering
Center for Advanced Technology for Large Structural Systems (ATLSS Center)
Lehigh University
Bethlehem, PA, USA

Volume 6

Structural Identification and Damage Detection using Genetic Algorithms

Chan Ghee Koh and Michael John Perry

Department of Civil Engineering, National University of Singapore

CRC Press
Taylor & Francis Group
Boca Raton   London   New York   Leiden

CRC Press is an imprint of the
Taylor & Francis Group, an **informa** business

A BALKEMA BOOK

Colophon

Book Series Editor:
Dan M. Frangopol

Volume Authors:
Chan Ghee Koh and Michael John Perry

Cover illustration:
Example of three buildings connected by two link bridges for output-only identification

Taylor & Francis is an imprint of the Taylor & Francis Group,
an informa business

© 2010 Taylor & Francis Group, London, UK

Typeset by Macmillan Publishing Solutions, Chennai, India
Printed and bound in Great Britain by Antony Rowe (a CPI Group company),
Chippenham, Wiltshire

British Library Cataloguing in Publication Data
A catalogue record for this book is available from the British Library

Library of Congress Cataloging-in-Publication Data

Koh, Chan Ghee.
  Structural identification and damage detection using genetic
algorithms / C.G. Koh and M.J. Perry.
     p. cm. -- (Structures and infrastructures series, ISSN 1747-7735; v. 6)
  Includes bibliographical references and index.
  ISBN 978-0-415-46102-3 (hardcover : alk. paper) — ISBN
978-0-203-85943-8 (e-book)   1. Structural analysis
(Engineering) — Mathematics.   2. Fault location
(Engineering) — Mathematics.   3. Genetic algorithms.   I. Perry, M. J.
(Michael J.), 1981– II. Title.   III. Series.

     TA646.K56   2010
     624.1'710151962—dc22

                                                                    2009038174

Published by: CRC Press/Balkema
                P.O. Box 447, 2300 AK Leiden, The Netherlands
                e-mail: Pub.NL@taylorandfrancis.com
                www.crcpress.com – www.taylorandfrancis.co.uk – www.balkema.nl

ISBN13   978-0-415-46102-3
ISBN13   978-0-203-85943-8 (eBook)
Structures and Infrastructures Series: ISSN 1747-7735
Volume 6

Table of Contents

Editorial

Welcome to the Book Series *Structures and Infrastructures*.

Our knowledge to model, analyze, design, maintain, manage and predict the life-cycle performance of structures and infrastructures is continually growing. However, the complexity of these systems continues to increase and an integrated approach is necessary to understand the effect of technological, environmental, economical, social and political interactions on the life-cycle performance of engineering structures and infrastructures. In order to accomplish this, methods have to be developed to systematically analyze structure and infrastructure systems, and models have to be formulated for evaluating and comparing the risks and benefits associated with various alternatives. We must maximize the life-cycle benefits of these systems to serve the needs of our society by selecting the best balance of the safety, economy and sustainability requirements despite imperfect information and knowledge.

In recognition of the need for such methods and models, the aim of this Book Series is to present research, developments, and applications written by experts on the most advanced technologies for analyzing, predicting and optimizing the performance of structures and infrastructures such as buildings, bridges, dams, underground construction, offshore platforms, pipelines, naval vessels, ocean structures, nuclear power plants, and also airplanes, aerospace and automotive structures.

The scope of this Book Series covers the entire spectrum of structures and infrastructures. Thus it includes, but is not restricted to, mathematical modeling, computer and experimental methods, practical applications in the areas of assessment and evaluation, construction and design for durability, decision making, deterioration modeling and aging, failure analysis, field testing, structural health monitoring, financial planning, inspection and diagnostics, life-cycle analysis and prediction, loads, maintenance strategies, management systems, nondestructive testing, optimization of maintenance and management, specifications and codes, structural safety and reliability, system analysis, time-dependent performance, rehabilitation, repair, replacement, reliability and risk management, service life prediction, strengthening and whole life costing.

This Book Series is intended for an audience of researchers, practitioners, and students world-wide with a background in civil, aerospace, mechanical, marine and automotive engineering, as well as people working in infrastructure maintenance, monitoring, management and cost analysis of structures and infrastructures. Some volumes are monographs defining the current state of the art and/or practice in the field, and some are textbooks to be used in undergraduate (mostly seniors), graduate and postgraduate courses. This Book Series is affiliated to *Structure and Infrastructure Engineering* (http://www.informaworld.com/sie), an international peer-reviewed journal which is included in the Science Citation Index.

It is now up to you, authors, editors, and readers, to make *Structures and Infrastructures* a success.

Dan M. Frangopol
Book Series Editor

About the Book Series Editor

Dr. Dan M. Frangopol is the first holder of the Fazlur R. Khan Endowed Chair of Structural Engineering and Architecture at Lehigh University, Bethlehem, Pennsylvania, USA, and a Professor in the Department of Civil and Environmental Engineering at Lehigh University. He is also an Emeritus Professor of Civil Engineering at the University of Colorado at Boulder, USA, where he taught for more than two decades (1983–2006). Before joining the University of Colorado, he worked for four years (1979–1983) in structural design with A. Lipski Consulting Engineers in Brussels, Belgium. In 1976, he received his doctorate in Applied Sciences from the University of Liège, Belgium, and holds two honorary doctorates (Doctor Honoris Causa) from the Technical University of Civil Engineering in Bucharest, Romania, and the University of Liège, Belgium. He is an Honorary Professor at Tongji University and a Visiting Chair Professor at the National Taiwan University of Science and Technology. He is a Fellow of the American Society of Civil Engineers (ASCE), American Concrete Institute (ACI), International Association for Bridge and Structural Engineering (IABSE), and the International Society for Health Monitoring of Intelligent Infrastructures (ISI IMII). He is also an Honorary Member of both the Romanian Academy of Technical Sciences and the Portuguese Association for Bridge Maintenance and Safety. He is the initiator and organizer of the Fazlur R. Khan Lecture Series (www.lehigh.edu/frkseries) at Lehigh University.

Dan Frangopol is an experienced researcher and consultant to industry and government agencies, both nationally and abroad. His main areas of expertise are structural reliability, structural optimization, bridge engineering, and life-cycle analysis, design, maintenance, monitoring, and management of structures and infrastructures. He is the Founding President of the International Association for Bridge Maintenance and Safety (IABMAS, www.iabmas.org) and of the International Association for Life-Cycle Civil Engineering (IALCCE, www.ialcce.org), and Past Director of the Consortium on Advanced Life-Cycle Engineering for Sustainable Civil Environments (COALESCE). He is also the Chair of the Executive Board of the International Association for Structural Safety and Reliability (IASSAR, www.columbia.edu/cu/civileng/iassar) and the Vice-President of the International Society for Health Monitoring of Intelligent Infrastructures (ISHMII, www.ishmii.org). Dan Frangopol is the recipient of several prestigious awards including the 2008 IALCCE Senior Award, the 2007 ASCE Ernest

Howard Award, the 2006 IABSE OPAC Award, the 2006 Elsevier Munro Prize, the 2006 T. Y. Lin Medal, the 2005 ASCE Nathan M. Newmark Medal, the 2004 Kajima Research Award, the 2003 ASCE Moisseiff Award, the 2002 JSPS Fellowship Award for Research in Japan, the 2001 ASCE J. James R. Croes Medal, the 2001 IASSAR Research Prize, the 1998 and 2004 ASCE State-of-the-Art of Civil Engineering Award, and the 1996 Distinguished Probabilistic Methods Educator Award of the Society of Automotive Engineers (SAE).

Dan Frangopol is the Founding Editor-in-Chief of *Structure and Infrastructure Engineering* (Taylor & Francis, www.informaworld.com/sie) an international peer-reviewed journal, which is included in the Science Citation Index. This journal is dedicated to recent advances in maintenance, management, and life-cycle performance of a wide range of structures and infrastructures. He is the author or co-author of over 400 refereed publications, and co-author, editor or co-editor of more than 20 books published by ASCE, Balkema, CIMNE, CRC Press, Elsevier, McGraw-Hill, Taylor & Francis, and Thomas Telford and an editorial board member of several international journals. Additionally, he has chaired and organized several national and international structural engineering conferences and workshops. Dan Frangopol has supervised over 70 Ph.D. and M.Sc. students. Many of his former students are professors at major universities in the United States, Asia, Europe, and South America, and several are prominent in professional practice and research laboratories.

For additional information on Dan M. Frangopol's activities, please visit www.lehigh.edu/~dmf206/

Preface

Structural health monitoring has become a growing R&D area, as witnessed by the increasing number of relevant journal and conference papers. Rapid advances in instrumentation and computational capabilities have led to a new generation of sensors, data communication devices and signal processing software for structural health monitoring. To this end, a crucial challenge is the development of robust and efficient structural identification methods that can be used to identify key parameters and hence, cause change of structural state. There are currently many competing methods of structural identification, both classical and non-classical. Based on our resarch efforts for over more than a decade, the genetic algorithms (GA) have been found to possess many desired characteristics and offer a very promising way to tackle real systems. It is the intention of this book, believed to be the first on this topic, to provide readers with the background and recent developments on GA-based methods for parameter identification, model updating and damage detection of structural dynamic systems.

Of significance, a novel identification strategy is developed which contains many advantageous features compared to previous studies. The application of the strategy focuses on structural identification problems with limited and noise contaminated measurements. Identification of systems with known mass is first presented to provide physical insight into the effects of various numerical parameters on the identification accuracy. Generalisation is then made to systems with unknown mass, stiffness and damping properties – a much tougher problem rarely considered in many other identification methods, due to the limitation of formulation in separating the effects of mass and stiffness properties.

The GA identification strategy is extended to structural damage detection whereby the undamaged state of the structure is first identified and used to direct the search for parameters of the damaged structure. Furthermore, another rarely studied problem of structural identification without measurement of input forces, i.e. output-only identification, is addressed which will be useful in cases where force measurement is difficult or impossible. It is our strong belief that any research attempt on structural identification and damage detection should be tested not only numerically but also experimentally, and hence a relatively long chapter on experimental study to validate the GA-based identification strategy. Finally, a practical divide-and-conquer approach of substructuring is presented to tackle large structural systems and also to illustrate the power and versatility of the GA-based strategy. The findings presented signify a quantum leap forward from research and practical viewpoints, and this book should therefore be useful to researchers, engineers and graduate students with interests in model updates, parameter identification and damage detection of structural and mechanical systems.

The authors wish to thank the staff of the Structural Engineering Laboratory of the Department of Civil Engineering at the National University of Singapore for their invaluable assistance in making the experimental study a success. The finanical support, including research scholarship for graduate students (including the second author) from the National University of Singapore is most appreciated. Many former and current graduate students, whose works have provided the foundation of this book in one way or another are also gratefully acknowledged. Special thanks go to Mr. Zhang Zhen and Mr. Trinh Ngoc Thanh for their great contributions and many insightful discussions.

Dedication

Chan Ghee **Koh** would like to dedicate this book to his family, Hwee Eng, Li Jia, Jessica Li Jian and Li Chen,

Michael John **Perry** would like to dedicate this book to his wife, Evelyn.

About the Authors

Chan Ghee Koh – Professor C.G. Koh received his PhD from the University of California, Berkeley, in 1986. His main research areas are structural dynamics, structural health monitoring and system identification. He has published more than 150 articles including more than 70 refereed international journal papers. He is a recipient of the prestigious Marie Curie Fellowship (1994) awarded by the Commission of the European Communities and of the IES Prestigious Publication Award (Best Paper in Theory, 1996) by the Institution of Engineers, Singapore. He was invited to deliver more than ten keynotes and invited lectures in U.K., Japan, China, India, Portugal and Greece. He is currently an associate editor of the International Journal on Structural Health Monitoring, and an editorial board member of the Journal of Smart Structures and Systems, as well as of the Journal of Vibroengineering.

Michael John Perry – Dr M. J. Perry gradated from the National University of Singapore with first class honours in civil engineering in 2003, under the Asia New Zealand Singapore Scholarship program. After receiving the award for the best civil engineering student, he continued his studies under NUS research scholarship and received his PhD in 2007. During his graduate study, he was one of the very few two-time recipients of the prestigious President Graduate Fellowship of NUS. While at NUS, his research focused on developing genetic algorithm identification strategies for structural and offshore applications. He is a co-author of two keynote papers and of a book chapter. Currently, he is a research engineer at Keppel Offshore & Marine Technology Centre, based in Singapore.

Chapter 1

Introduction

Buildings, bridges, offshore platforms, dams and other civil infrastructures may experience damage during their service life due to natural and man-made actions. Significant damage in a structure is often manifested through changes in physical properties, such as decrease in structural stiffness and a corresponding shift of natural frequencies. If not monitored and rectified early, damage would compromise the performance of structure, increase maintenance cost and, in the unfortunate event, result in structural failure. From the viewpoint of functionality and safety, it is therefore essential and beneficial to have means of early detection of structural damage. To this end, structural damage identification has now become a vital component of an emerging engineering discipline known as Structural Health Monitoring (SHM). Applicable to civil infrastructures as well as mechanical, aerospace and other types of structures, SHM involves the observation of structures by measurement to determine the "health" or "fitness" of structures under gradual or sudden changes to their state. Some of the recent noteworthy efforts in SHM are reported in special issues in journals such as *Journal of Engineering Mechanics*, ASCE (Ghanem and Sture, 2000; Bernal and Beck, 2004), *Computer-Aided Civil and Infrastructure Engineering* (Adeli, 2001), *Smart Materials and Structures* (Wu and Fujino, 2006), *Structure and Infrastructure Engineering* (Chang, 2007), and *Philosophical Transactions of the Royal Society A* (Farrar and Worden, 2007).

The rapidly growing interest in SHM can be partly attributed to technological advances in sensors, data acquisition and processing, wireless communication, etc, and partly attributed to the rising awareness of its long-term benefits by the owners, operators and authorities. Tangible benefits of SHM to the users include better performance prediction, lower life-cycle cost and more reliable evaluation of structural safety (see, for example, Frangopol and Messervey, 2009; Liu et al., 2009a and b). With increasing acceptance of response monitoring, the need for more efficient and robust algorithms to extract useful information from the enormous data collected is more than ever. For the purpose of structural identification and damage detection, the use of dynamic response is usually preferred over static response as dynamic signals offer more information and avoid the possible non-uniqueness problem. For static methods to overcome the non-uniqueness problem, force application at multiple locations is usually required (Sanayei and Onipede, 1991; Hjelmstad and Shin, 1997) or additional information such as modal frequencies are needed (Wang et al., 2001). Noise effect in dynamic measurement can normally be filtered out by low-pass filters

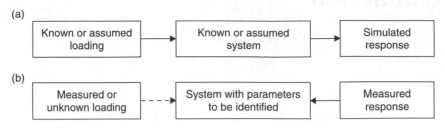

Figure 1.1 (a) Direct analysis (simulation); (b) inverse analysis (identification).

unlike in static measurement. Furthermore, it is easier to carry out dynamic measurement via accelerometers than static measurement. Static measurement requires a fixed reference for the displacement sensors or incurs numerical error if integrated twice from accelerometer signals. For these practical reasons, most of research works on structural identification and damage detection have been based on dynamic measurement, in which case the methods may be referred to as vibration-based identification or vibration-based damage detection. This normally involves inverse analysis which is more difficult to do than forward analysis. In forward analysis, the aim is predict the response (output) for given excitation (input) and known system parameters. Inverse analysis dealing with identification of system parameters based on given input and output (I/O) information (Fig. 1.1) is known as system identification. If only output information is needed, this is known as output-only system identification.

1.1 Modelling and Simulation of Dynamic Systems

System identification, in a broad sense, can be described as the identification of the conditions and properties of mathematical models that aspire to represent real phenomena in an adequate manner. Originally used in electrical and control engineering and subsequently extended to the fields of mechanical, aerospace and civil engineering, system identification typically involves the following two key aspects:

- Choosing a mathematical model that is characterized by a finite set of key parameters
- Identifying these parameters based on measurement signals

The success of damage identification hinges on, to a large extent, realistic modelling of the structural system as well as efficient numerical simulation to obtain the dynamic response. While it is appealing to adopt a detailed structural model, for example, by means of the finite element method, the resulting number of degrees of freedom (DOFs) for a real structure would usually be very large. This would translate into higher computational cost and, if the number of unknowns is also large, greater difficulty in system identification. Hence, special attention is needed to keep the size of structural model sufficiently representative of the main features of the real structure and at the same time keep the computational effort at an affordable level. Often, what is needed for damage detection is to detect changes of key structural parameters, for instance, the storey stiffness values for which a shear building model or even a lumped mass model may suffice. In this regard, the concept of substructural identification is very attractive because it reduces the system size so that parameter identification is executed with a manageable number of unknowns. This will be illustrated in Chapter 8 of the book.

In a broader sense, modelling error includes error arising from the numerical scheme, if required, in "integrating" the equations of motion. This error source can be reduced by using a higher-order numerical scheme or a small time step, either of which would entail higher computational cost. Hence the issue of duration of time signals considered is important for the efficiency of the identification strategy. Depending on the algorithm used, it may not be worthwhile to use a long duration of time signals in the early stage of identification; this will be addressed in Chapter 3.

Recognizing the fact that modelling error exists no matter how refined the numerical model (including integration scheme) is used, a preferred strategy is to focus on damage detection by comparing the changes of the monitored state and the reference state. The reference state is usually the undamaged state, i.e. when the structure is new or deemed to be free of any significant damage. By including the undamaged state in the system identification and using its results as the benchmark, the model error can be reduced – provided that the model is sufficiently accurate. This forms the basis of structural damage detection as discussed in Chapter 6.

1.2 Structural Identification and Damage Detection

The identification of stiffness, mass and/or damping of a structural system is referred to herein as "structural identification" in short. Structural identification can be used to update or calibrate structural models so as to better predict response and achieve more cost-effective designs. More importantly, by tracking changes of key parameters, structural identification can be used for non-destructive assessment due to damaging events such as earthquakes and also for deterioration monitoring of ageing structures over time. There are three important components to damage detection, in the order of difficulty, as follows: (1) damage alarming, i.e. to indicate whether there is damage; (2) damage localization, i.e., to identify the location of damage; and (3) damage quantification, i.e. to quantify the extent of damage.

From a computational point of view, identification of a dynamic system can be a daunting task, particularly when the system involves a large number of unknown parameters. The effectiveness of an identification strategy can be measured in terms of accuracy, efficiency and robustness. Robustness in this context refers to the high success rate of finding the solution with as little requirements as possible in terms of, for instance, initial guess and gradient information.

1.3 Overview of Structural Identification Methods

Many different methods have been developed for structural identification; they are too numerous to be given a thorough review here. Recent literature reviews of structural identification from different perspectives can be found in Chang et al. (2003), Carden and Fanning (2004), Hsieh et al. (2006), Humar et al. (2006) and Friswell (2007).

When system identification is treated as an optimization problem in terms of minimizing the errors between the measured and predicted signals, the methods can be categorized as classical and non-classical methods. Classical methods are typically those derived from sound mathematical theories. They perform point-to-point search and often require the gradient information (or its variant) to guide its search direction. Depending largely on the initial guess, the solutions may converge falsely to a local optimal point rather than the global optimum. The classical methods can be categorized according to whether the identification is carried out directly from the measured time signals or from the frequency domain information via Fourier transform. Some of the commonly adopted classical methods are introduced in the following sections, first in the frequency domain and then in the time domain.

1.3.1 *Frequency Domain Methods*

Identification of dynamic properties and damage in the frequency domain is based mainly on measured natural frequencies and mode shapes. Time signals are digitally converted to extract these modal properties by fast Fourier transform (FFT) (Cooley and Tukey, 1965) or similar algorithms. Loss of stiffness, representing damage to the structure, is detected when measured natural frequencies are significantly lower than expected. A useful review on the use of frequencies in detecting structural damage is given in Salawu (1997).

There has been substantial discussion as to the change in frequency required to detect damage, and also if changes in frequencies due to environmental effects can be separated from those due to damage. Creed (1987) estimated that it would be necessary for a natural frequency to change by 5% for damage to be confidently detected. Case studies on an offshore jacket and a motorway bridge showed that changes of frequency in the order of 1% and 2.5% occurred due to day to day changes in deck mass and temperature respectively. Numerical simulation studies showed that large damage, for example from the complete loss of a major member would be needed to achieve the desired 5% change in frequencies. Aktan et al. (1994) suggested that frequency changes alone do not automatically suggest damage. They reported frequency shifts for both steel and concrete bridges exceeding 5% due to changes in ambient conditions within a single day. They also reported that the maximum change in the first 20 frequencies of a RC slab bridge was less than 5% after it had yielded under an extreme static load. More recently, Catbas et al. (2008) demonstrated the significant effect of environmental conditions (particularly the temperature) on the reliability estimation through the SHM study of a long span truss bridge.

Notwithstanding the above findings, some researchers claimed success using natural frequencies. For example, Adams et al. (1978) reported very good success in detecting damage in relatively simple one-dimensional structures. Small saw cuts were identified and located using changes in the first 3 natural frequencies for simple bars, tapered bars and a cam shaft. The limitation of the study was the need of highly accurate frequency measurements to six significant digits. In addition, the location of damage could only be obtained if at least $2n$ frequencies were available, where n is the number of damage locations.

Identification can also be carried out using criteria based on mode shapes. These methods can be based on a direct comparison of displacement mode shapes or curvature mode shapes. Two methods are commonly used for direct comparison of mode shapes. The modal assurance criterion (MAC) indicates correlation between two sets of mode shapes while the coordinate modal assurance criterion (COMAC) indicates the correlation between mode shapes at selected points on the structure. As the greatest change in mode shapes is expected to occur at the damage location, COMAC can be used to determine the approximate location of damage. **MAC** is defined as shown in equation 1.1 whereby $\mathbf{\Phi}_u$ and $\mathbf{\Phi}_d$ are the mode shape matrices obtained for the undamaged structure (denoted by subscript u) and for the damaged structure (denoted by subscript d). If the structure is undamaged **MAC** becomes an identity matrix. The COMAC is computed for a given point (j) by summing the contributions of n modes as shown in equation 1.2. The COMAC value should be one for undamaged location and less than one if damage is present.

$$\mathbf{MAC} = \frac{(\mathbf{\Phi}_u^T \mathbf{\Phi}_d)^2}{(\mathbf{\Phi}_u^T \mathbf{\Phi}_u)(\mathbf{\Phi}_d^T \mathbf{\Phi}_d)} \tag{1.1}$$

$$COMAC(j) = \frac{\sum_{i=1}^{n} (\varphi_{u,ij} \varphi_{d,ij})^2}{\sum_{i=1}^{n} (\varphi_{u,ij} \varphi_{u,ij}) \sum_{i=1}^{n} (\varphi_{d,ij} \varphi_{d,ij})} \tag{1.2}$$

Salawu and Williams (1995) conducted full scale tests on a reinforced concrete highway bridge before and after repairs were carried out. Their results showed that, while natural frequencies varied by less than 3%, the diagonal MAC values ranged from 0.73 to 0.92 indicating a difference in the state of the structure. Using a threshold level of 0.8 the COMAC values were able to locate damage at 2 of 3 damaged locations, but also identified damage at 2 undamaged locations. Fryba and Pirner (2001) used the COMAC criteria to check the quality of repairs carried out to a concrete bridge which had slid from its bearings. The modes of the undamaged and repaired halves of the building were compared to demonstrate that the repairs had been well done. Mangal et al. (2001) conducted a series of impact and relaxation tests on a model of an offshore jacket. They found that significant changes in the structural modes occurred for damage of critical members as long as they were aligned in the direction of loading. The relaxation type loading gave results as good as the impact loading indicating it to be a good alternative for future studies.

The use of mode shape curvature in damage detection assumes that changes in curvature of mode shapes are highly localised to the region of damage and are more sensitive to damage than the corresponding changes in the mode shapes. Wahab and De Roeck (1999) used changes in modal curvature to detect damage in a concrete bridge. The modal curvature was computed from central difference approximation and a curvature damage factor (CDF) used to combine the changes in curvature over a number of modes. The method was able to identify the damage location but only for the largest damage case tested.

While much effort has gone into developing the frequency and mode shape methods, as mentioned above, significant doubt still remains as to the sensitivity of the tests to realistic levels of damage. To address this problem, other methods that are claimed to be more sensitive to damage have been developed. The flexibility of a structure is the inverse of its stiffness and may be estimated from the measured frequencies (ω) and modes ($\mathbf{\Phi}$) as shown in equation 1.3 (Raghavandrachar and Aktan, 1992). Typically, not all modes of a structure can be measured. Nevertheless, a reasonable estimate of the flexibility is obtained using a limited number of modes. Studies carried out by Aktan et al. (1994) and Zhao and DeWolf (1999) showed that for structural damage detection, modal flexibilities could give a better indication of damage than the measured frequencies or mode shapes alone.

$$\mathbf{F} = \mathbf{\Phi} \frac{1}{\omega^2} \mathbf{\Phi}^T \tag{1.3}$$

A comparison of the performance of several methods is provided in Farrar and Doebling (1997). A study of various levels of damage on the I-40 bridge over the Rio Grande was identified using changes in modes, mode shape curvature, flexibility, stiffness and a damage index method (e.g. Kim and Stubbs, 1995). The study showed the damage index method to give the best results while the flexibility method failed on all but the largest damage case.

An advantage of the frequency domain methods is that the input force measurement may not be required. In fact, input characteristics may also be identified along with the system parameters. Shi et al. (2000) applied a filter method to the frequency domain to identify system and input parameters for both simulated and experimental examples. Spanos and Lu (1995) introduced a decoupling method in frequency domain to identify the structural properties and force transfer parameters for the non-linear interaction problems encountered in offshore structural analysis. Roberts and Vasta (2000) used standard second order spectra and higher order spectra to simultaneously estimate the system and excitation process parameters from the measured response.

1.3.2 Time Domain Methods

A major drawback of frequency based methods is that for real structures information for higher modes of vibration will be unreliable due to low signal to noise ratio. In addition the methods usually involve modal superposition limiting the application to linear systems. Finally, frequencies are a global property and are rather insensitive to local damage. Identifying and locating damage is therefore very difficult, particularly when only the first few modes of vibration can be measured. Time domain methods remove the need to extract frequencies and modes and, instead, make use of the dynamic time-history information directly. In this way information from all modelled modes of vibration are directly included. In addition, non-linear models can be identified as there is no requirement for the signal to be resolved into linear components. Ljung and Glover (1981) noted that while frequency and time domain methods should be viewed as complementary rather than rivalling, if prior knowledge of the system is available and a model to simulate time-histories is to be obtained, time domain methods should be adopted. The more established classical time-domain methods include least squares method, instrumental variable method, maximum likelihood method, extended Kalman filter method, observer Kalman filter identification method, Monte Carlo filter method and eigensystem realization algorithm. Some of these methods are discussed as follows.

1.3.2.1 Least Squares Method

The least squares (LS) method was one of the earliest classical identification techniques in time domain. The method works by minimising the sum of squared errors between the measured response and that predicted by the mathematical model. As an illustration example, consider the case of a single-degree-of-freedom forced oscillation which may be modelled as

$$m\ddot{x} + c\dot{x} + kx = F \tag{1.4}$$

where x, $\dot{x}$ and $\ddot{x}$ are the displacement, velocity and acceleration of the oscillator caused by the excitation force F. The least squares method can be used to solve for the mass m, stiffness k and damping c of the oscillator by minimising the error in the force estimated from the measured response of the structure using the structural model. The method assumes the inputs to be correct and error to occur only as output noise. At a given time step the measured force F_k is therefore the sum of the estimated force $\hat{F}_k$ and an output error ε_k as

$$F_k = \hat{F}_k + \varepsilon_k = m\ddot{x}_k + c\dot{x}_k + kx_k + \varepsilon_k \tag{1.5}$$

or in standard form as follows:

$$y_k = \hat{y}_k + \varepsilon_k = \varphi_k\theta + \varepsilon_k \tag{1.6}$$

where the output y, regressor φ_k, and parameter vector θ, represent the force F, response $[\ddot{x}_k\ \dot{x}_k\ x_k]$ and parameters $[m\ c\ k]^T$ of the system, respectively. With N data points available the output and regressor can form matrices with N rows as

$$\mathbf{Y} = \begin{bmatrix} y_1 \\ y_2 \\ \vdots \\ y_N \end{bmatrix} \qquad \boldsymbol{\Phi} = \begin{bmatrix} \varphi_1 \\ \varphi_2 \\ \vdots \\ \varphi_N \end{bmatrix} \qquad \hat{\mathbf{Y}} = \begin{bmatrix} \hat{y}_1 \\ \hat{y}_2 \\ \vdots \\ \hat{y}_N \end{bmatrix} = \boldsymbol{\Phi}\theta \tag{1.7}$$

The output error is assumed to be a random Gaussian variable with zero mean. The least squares method identifies estimates for the parameters, $\hat{\theta}$ by minimising the sum of squared errors (SSE) between the measured and estimated output.

$$SSE = \sum_{k=1}^{N} (y_k - \varphi_k\hat{\theta})^2 \tag{1.8}$$

The above error is minimised by setting the derivative to zero.

$$0 = \frac{d}{d\hat{\theta}} SSE(\theta) = \frac{1}{2} \sum_{k=1}^{N} \varphi_k^T (y_k - \varphi_k\hat{\theta})$$

$$\sum_{k=1}^{N} \varphi_k^T y_k = \sum_{k=1}^{N} \varphi_k^T \varphi_k\hat{\theta} \tag{1.9}$$

$$\hat{\theta} = \left[\sum_{k=1}^{N} \varphi_k^T \varphi_k\right]^{-1} \sum_{k=1}^{N} \varphi_k^T y_k$$

This leads to the well known least squares estimate for θ.

$$\hat{\theta} = [\Phi^T \Phi]^{-1} \Phi^T Y \tag{1.10}$$

It should be noted that, while the force is used as the output of the system in the example above, this does not have to be the case. For example, the displacement can be used as output by rearranging the equation of motion as;

$$x = \frac{F}{k} - \frac{m}{k}\ddot{x} - \frac{c}{k}\dot{x} \tag{1.11}$$

The regressor and parameter vectors would now become

$$\varphi_k = [F_k \;\; \ddot{x}_k \;\; \dot{x}_k] \qquad \theta = \left[\frac{1}{k} \;\; \frac{m}{k} \;\; \frac{c}{k}\right]^T \tag{1.12}$$

The mass, stiffness and damping parameters are not directly identified, but can easily be extracted from the estimated parameters. In many previous studies it is assumed that the mass is known and thus the inertia term ($m\ddot{x}$) is grouped with the force reducing the problem to two unknowns.

While the LS method has a good mathematical basis, it has difficulty when dealing with real data as noise and inadequacy of system models can cause the results to deviate significantly. Though the derivation of the method assumes noise on the output, it does not allow for noise in the regressor, which is unavoidable in a real situation. The method also requires full measurement of the system, rendering it nonviable for large systems with many DOFs.

As one of the first time domain methods applied to structural identification problems, the LS method has received a good deal of attention. Caravani et al. (1977) developed a recursive algorithm for computing the least squares estimate without matrix inversion and applied it to the identification of a 2-DOF shear building. An interesting iterative method was proposed by Ling and Haldar (2004). They used a least squares method with iteration to identify structural properties without using any input force information. The method worked by alternating between identification of parameters, using an assumed force, and then updating the force using the identified parameters. By using several iterations of this procedure the parameters and applied forces could be identified. The method was demonstrated on several example problems using both viscous and proportional damping models. Identification of structural parameters in the time domain without the need for force measurement is a very promising direction. This idea is explored further with a new output-only identification method in Chapter 5 of this book.

1.3.2.2 Instrumental Variable Method

This method is similar to the recursive least-square method, in the sense that square-error norm between the estimated and measured responses is minimised. The equation of the response forecast is same as Eq. (1.6). A vector of instrumental variables (ξ) which is highly correlated with φ but uncorrelated with the prediction error e is introduced into the criterion function. Unknown parameters are also updated by setting the gradient of the criterion function with respect to the unknown parameters to zero (Imai et al. 1989). The instrumental variable estimation is given by

$$\hat{\theta}_k^{iv} = \left[\frac{1}{k} \sum_{i=1}^{k} \xi(i) \varphi^T(i) \right]^{-1} \frac{1}{k} \sum_{=1}^{k} \xi(i) \mathbf{y}(i) \tag{1.13}$$

This method can handle measurement with noise. However, a good initial guess is still required for this method to work well.

1.3.2.3 Maximum Likelihood Method

As measured responses are often contaminated by noise, which is usually random in nature, the identified parameters should be treated as random variables. It is therefore justifiable to determine unknown parameters by maximising the likelihood (probability density function) of matching the estimated responses with the measured responses. This is known as the maximum likelihood method (Yun and Shinozuka, 1980; Shinozuka et al., 1982; DiPasquale and Cakmak, 1988; and Ljung, 1986). This method has the advantage of providing the best estimation for a wide range of contamination intensity in the excitation force and the structural response.

In maximizing the likelihood function, it is more convenient to take the logarithm. Since the logarithm is monotonic, the transformation does not change the optimal point. The likelihood function can be written in the following form

$$\mathbf{L}(\theta, \varepsilon_i) = const - \frac{1}{2} \log \det \Lambda_i(\theta) - \frac{1}{2} \varepsilon_i^T \Lambda_i^{-1}(\theta) \varepsilon_i \tag{1.14}$$

where θ = vector of unknown parameters, $\Lambda_i(\theta)$ = covariance matrix of prediction errors (ε). The maximum likelihood method has been proven to have superior convergence properties over the least-square method. However, it usually requires a larger amount of computational time. Derivatives are also required in this method. Furthermore, the optimization process is relatively sensitive to the initial guess used.

1.3.2.4 Kalman Filter Method

Some of the most commonly used time domain methods today are modifications of the Kalman filter (Kalman, 1960). The Kalman filter is a set of mathematical equations that provides a recursive means to estimate the state of a process in a way that minimises the mean of the square error. An introduction to the Kalman filter can be found in Welch and Bishop (2004) and Maybeck (1979). The filter estimates the state $\mathbf{x}$, of a discrete time process governed by the linear stochastic difference equation with input $\mathbf{u}$, and measurement $\mathbf{z}$, which is related to the state by observation equation. The system matrices $\mathbf{A}$ and $\mathbf{B}$ relates the current state to the previous state and the system inputs while the matrix $\mathbf{H}$ relates the measurement to the state of the system. The process and measurement noise ($\mathbf{w}$ and $\mathbf{v}$ respectively) are assumed to be zero mean Gaussian noise with covariances of $\mathbf{Q}$ and $\mathbf{R}$ respectively, that is, $\mathbf{w} \sim N(0,\mathbf{Q})$ and $\mathbf{v} \sim N(0,\mathbf{R})$.

$$\mathbf{x}_k = \mathbf{A}\mathbf{x}_{k-1} + \mathbf{B}\mathbf{u}_{k-1} + \mathbf{w}_{k-1} \tag{1.15}$$

$$\mathbf{z}_k = \mathbf{H}x_k + \mathbf{v}_{k-1} \tag{1.16}$$

The Kalman filter can be thought of in terms of a predictor step followed by a corrector step. The predictor step is used to find an estimate of $\mathbf{x}$ at time step k from the knowledge of the process prior to k. This estimate, denoted $\hat{\mathbf{x}}_k^-$, is estimated from equation 1.15 assuming the noise term is zero. The corrected state $\hat{\mathbf{x}}_k$, is then obtained as a weighted combination of the predicted state and the state obtained from the measured response as follows

$$\hat{\mathbf{x}}_k = \hat{\mathbf{x}}_k^- + \mathbf{K}_k(\mathbf{z}_k - \mathbf{H}\hat{\mathbf{x}}_k^-) \tag{1.17}$$

The errors of the predicted and corrected states are therefore

$$\mathbf{e}_k^- = \mathbf{x}_k - \hat{\mathbf{x}}_k^- \tag{1.18}$$

$$\mathbf{e}_k = \mathbf{x}_k - \hat{\mathbf{x}}_k \tag{1.19}$$

The error covariance for the predicted and corrected states are estimated as

$$P_k^- = E[\mathbf{e}_k^- \mathbf{e}_k^{-T}] \tag{1.20}$$

$$P_k = E[\mathbf{e}_k \mathbf{e}_k^T] \tag{1.21}$$

The Kalman gain $\mathbf{K}$, is selected to minimise the error covariance of the estimated state. One form of $\mathbf{K}$ which minimises the error covariance (1.21) is shown below:

$$\mathbf{K}_k = P_k^- \mathbf{H}^T [\mathbf{H} P_k^- \mathbf{H}^T + \mathbf{R}]^{-1} \tag{1.22}$$

From this equation we can see that as the measurement error covariance $\mathbf{R}$ approaches zero, the gain approaches $\mathbf{H}^{-1}$ and the state estimate (Eq 1.17) is dominated by the measurement. On the other hand if the a priori estimate error covariance approaches zero, the gain becomes zero and the estimate is dominated by the predicted state. In effect the Kalman gain reflects how much we 'trust' the measured and predicted states. In practice the initial estimates of the state $\mathbf{x}_0$, error covariance $\mathbf{P}_0$, and noise covariances $\mathbf{R}$ and $\mathbf{Q}$ are needed to get the filter started. The choice of $\mathbf{P}_0$ is not critical as it will converge as the filter proceeds, while $\mathbf{R}$ and $\mathbf{Q}$ should be given reasonable values in order for the solution to converge. The Kalman filter is summarised in figure 1.2. The basic linear Kalman filter described above can also be linearized about the current operating point for use in non-linear systems. Referred to as the Extended Kalman Filter (EKF) this powerful modification has allowed for application of the filter into many identification and control problems.

For identification problems an augmented state vector containing the system state and the system parameters to be identified is used (Carmichael, 1979). The parameters are then estimated along with the state as the filter proceeds. Hoshiya and Saito (1984) proposed that several iterations of the EKF, with the error covariance weighted between iterations, could lead to more stable parameter estimation. The weighted global iteration procedure was demonstrated for 2- and 3-DOF linear and bilinear hysteretic systems. Koh and See (1994, 1999) proposed an adaptive EKF method which updates the system noise covariance in order to enforce consistency between residuals and their statistics. The method is able to estimate parameters as well as give a useful estimate of their uncertainty.

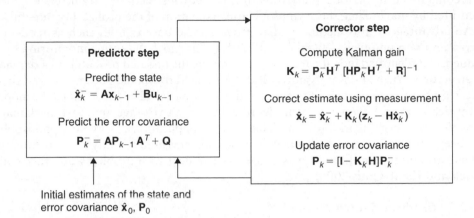

Predictor step

Predict the state

$$\hat{x}_k^- = Ax_{k-1} + Bu_{k-1}$$

Predict the error covariance

$$P_k^- = AP_{k-1}A^T + Q$$

Corrector step

Compute Kalman gain

$$K_k = P_k^- H^T [HP_k^- H^T + R]^{-1}$$

Correct estimate using measurement

$$\hat{x}_k = \hat{x}_k^- + K_k(z_k - H\hat{x}_k^-)$$

Update error covariance

$$P_k = [I - K_k H]P_k^-$$

Initial estimates of the state and error covariance $\hat{x}_0$, P_0

Figure 1.2 Kalman filter.

1.3.2.5 *Monte Carlo Filter Method*

Monte Carlo filter (MCF) was first proposed by Kitagawa (1996). Structural para-meters are derived by obtaining recursively the conditional distribution function of the state variable when observation values up to the present time step are given. The distribution function of a state vector is described by many samples instead of first and second moments, unlike in the case of EKF. The MCF has an advantage that it can deal with nonlinear and non-Gaussian noise problem. A modified approach called the adaptive MCF method was developed by Sato and Kaji (2000). This method identifies systems with rapidly changing parameters incorporating a "forgetting" factor to express the rate of diminishing effect of past observation data in the covariance of the adaptive noise. The adaptive noise, which is non-Gaussian and independent of state variables, is introduced in the state transfer equation to enlarge variance of the distribution of predictor. Hence, the identified structural parameters become much dependent on the recent data observed and the reliability of past observation data can be reduced. Yoshida and Sato (2002) proposed a method of damage detection using MCF. The formulation is a natural extension of Kalman filter (linear Gaussian) and does not necessarily require Gaussian noise. Nevertheless, the MCF requires many particles (samples) and hence a high computational cost in order to describe the detailed probabilistic nature of the identified parameters.

1.3.2.6 *Bayesian Method*

Beck and Katafygiotis (1998) presented a Bayesian statistical framework for system identification whereby probability models are used to account for parameter uncer-tainty and prediction uncertainty. Formulating the weighted probability models in the form of initial predictive probability density function, Bayes' theorem is applied to update the predictive PDF. Nevertheless, the initial predictive PDF for the system out-put is usually a multidimensional integral which is difficult to evaluate. This difficulty is overcome by an asymptotic approximation. Unknown structural parameters are then identified by maximizing the asymptotic approximation of the probability integral.

An advantage of this method is that it can handle uncertainties such as modeling errors and non-uniqueness. Vanik et al. (2000) simulated an on-line monitoring by a sequence of identified modal parameter to compute the updated probability of damage of structures. Yuen and Katafygiotis (2001) estimated the modal parameters and their uncertainties using only one set of ambient data. Yuen et al. (2004) combined the modal identification and Bayesian system identification in a two-stage approach in damage detection of a benchmark problem. Thus far, the application of Bayesian philosophy has been confined to small-scale identification problems until recently when an attempt by substructural identification was carried out to deal with problems of larger scale (Yuen and Katafygiotis 2006).

1.3.2.7 Gradient Search Methods

Some researchers have tackled structural identification problems by gradient search methods, for example, Gauss-Newton least square (Bicanic and Chen 1997; Chen and Bicanic 2000) and Newton's method (Liu and Chen 2002; Lee 2009). These methods have the drawbacks such as the need of good initial guess and gradient information (which can be difficult to obtain for structural identification problems). More importantly, these classical methods commonly lack global search capability and tend to converge prematurely to local optima. Hence, these methods tend to be ineffective in the presence of noise (Liu and Chen 2002).

1.3.3 Non-Classical Methods

Many of the classical methods discussed in the previous sections have limitations in one way or another. Some classical methods require gradient information to guide the search direction, which normally would require relatively good initial guess in order for the solution to converge. Some classical methods work on transformed dynamic models, such as state space models, where the identified parameters lack physical meaning. This may often make it difficult to extract and separate physical quantities such as mass and stiffness. The associated state space formulation would usually require time histories of displacement and velocity which, if integrated from measured acceleration, would incur numerical error. In addition, a recent trend of research is towards identification of large systems with as many unknown parameters as possible. For large systems, many classical methods suffer the ill-condition problem and the difficulty of convergence increases drastically with the number of unknown parameters.

To reduce the dependence on initial guess and increase the success rate of global search, exploration methods such as random search may be used but are obviously not efficient for large systems due to the huge combinatorial possibilities. Some heuristic rules are needed to define the search strategy and these rules are typically non-mathematical in nature leading to non-classical methods. These methods usually depend on computer power for an extensive and hopefully robust search. As computer power has rapidly increased in recent years, the use of heuristic-based non-classical methods has become very attractive. To date, the two main non-classical methods used for structural identification are genetic algorithms (GA) and neural network. The neural network method for structural identification will be briefly reviewed in the next section. The application of GA in civil engineering has also attracted tremendous interest from researchers and practitioners in recent years. For example, Furuta et al. (2006) adopted an improved multi-objective genetic algorithm to develop a bridge management system that can facilitate practical maintenance plan. The proposed cost-effective decision-support system was verified via the investigation on a group of bridges. Okasha and Frangopol (2009) incorporated redundancy in lifetime maintenance optimization based system reliability, and used GA to obtain solutions to the multi-objective optimization problem by considering system reliability, redundancy and life-cycle cost. The GA-based structural identification methods are the main focus of this book and its principles will be explained in detail in Chapter 2.

Recently, several other non-classical methods have also been reported. It is beyond the scope of this book to provide a comprehensive review as they are relatively new and still growing. Some representative examples are given here. For instance, evolutionary strategy was studied to identify 3-DOF and 10-DOF lumped systems (Franco et al. 2004). A differential evolution strategy was also investigated for identifying physical parameters in time domain (Tang et al. 2008). Fuzzy logic, coupled with principles of continuum damage mechanics, is used to identify the location and extent of structural damage (Sawyer and Rao, 2000). The proposed methodology represents a unique approach to damage detection that can be applied to a variety of structures used in civil engineering and machine and aerospace applications. Simulated annealing was combined with genetic algorithms to detect damage of beam structures via static displacement and natural frequencies (He and Hwang 2006). Particle swarm optimization (PSO) was used for structural identification due to its simple concept and quick convergence (Tang et al. 2007). PSO coupled with simplex algorithm was found to perform better than simulated annealing and basic PSO in damage identification using frequency domain data (Begambre and Laier 2009). Imitating the self-organization capability of ant colony, Li et al. (2006) proposed a biologically inspired search method to identify parameters of a chaotic system.

Collectively, these non-classical methods can also be called soft computing methods as they rely on (soft) heuristic concepts rather than (hard) mathematical principles. Due to their great potential in handling difficult problems (e.g. inverse problem as in the case of structural identification), there has been substantial increase in R&D interest as evident in the many papers presented in a recent conference on soft computing technology (Topping and Tsompanakis, 2009).

1.3.3.1 Neural Network Method

Neural network (NN) method has gained popularity as it is relatively easy to implement in discovering similarities when confronted with large bodies of data. NN is the functional imitation of a human brain and works by combining layers of 'neurons' through weighted links. At each neuron the weighted inputs are processed using some simple function to obtain the output from the neuron. A basic neural network usually contains 3 layers, an input layer, hidden layer and output layer as illustrated in figure 1.3. By correct weighting of the connections and simple functions at the neurons, the inputs can be fed through the network to arrive at the outputs for both linear and non-linear systems. The beauty of neural networks lies in the fact that they can be 'trained'. This means that through some process the network can adjust its weights to

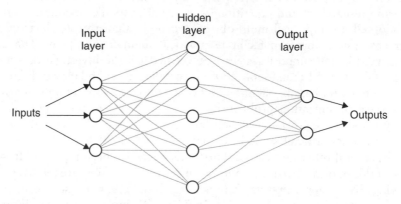

Figure 1.3 Layout of a simple neural network.

match given input/output sequences. This pattern recognition ability has allowed the application of neural networks to artificial intelligence applications.

Several training methods for neural networks have been developed, the most popular of which is the back propagation algorithm. This involves feeding the errors at the output layer back through the net to adjust the weights on each link. Other methods such as the probabilistic neural network have also been developed. An early example of the application of NN to system identification is given in Chen et al. (1990). They used multilayer neural networks for the identification of non-linear autoregressive moving average with exogenous inputs systems. Due to its strengths in pattern recognition and classification, NN has been used in structural identification and damage detection in recent years (Tsai and Hsu, 1999; Adeli and Karim, 2000; Ni et al., 2002; Yeung and Smith, 2005; Jiang et al., 2006).

For a SHM system, its efficiency is mainly determined by the diagnosis methods and data from numerous sensors of identical or dissimilar types. Sensors are needed extensively in SHM to provide sufficient input and output. For example, in a fairly comprehensive long-term monitoring system, Tsing Ma Bridge in Hong Kong is permanently instrumented with about 300 sensors of various types (Ko et al., 1999; Ko and Ni, 2005; Chan et al., 2006). The main drawback in the use of NN for large-scale system identification is that huge amount of data are required to properly train the network. A lack of some patterns of data will cause the identification to return incorrect values.

Chapter 2

A Primer to Genetic Algorithms

The identification strategies used throughout this book are based on genetic algorithms (GAs) which are inspired by Darwin's theory of natural selection and survival of the fittest. Darwin observed that individuals with characteristics better suited for survival in the given environment would be more likely to survive to reproduce and have their genes passed on to the next generations. Through mutations, natural selection and reproduction, species could evolve and adapt to changes in the environment. In a similar way it is possible to evolve solutions to a problem through mathematical operators which mimic the natural selection processes present in nature.

In this chapter an understanding of the functioning of genetic algorithms is developed. The ideas behind GA and how GA differs from other search algorithms are first established. The genetic operators are then described using an example problem and a basic mathematical theory is given to explain why GAs work, providing an insight into how random processes can be directed to search for the desired solutions. The orignial GA, in its earlier form, is suitable for simple problems such as the finding of maximum or minimum of a mathematical function. For more complex engineering problems, some limitations exist with the original GA. Some of the problems associated with the original GA have been overcome in recent times. The chapter concludes with a discussion of the recent advances and modifications that have been suggested in order to improve the performance of GAs.

2.1 Background to GA

The major early work on adaptation based on GA was by John H. Holland (1975) in his book: *Adaptation in Natural and Artificial Systems*. Adaptation is regarded as a process of progressive variation of structures, leading to an improved performance. He recognized the similarities between natural and artificial systems and sought ways in which the operators acting to shape the development of natural systems could be modelled mathematically. Recognising that operators such as crossing over and mutation that act in natural systems were also present in many artificial systems, Holland proposed that computers could be programmed by specifying 'what has to be done', rather than 'how to do it'.

GAs are search algorithms that combine a 'survival of the fittest' mentality with a structured, yet random, exchange of information in order to explore the search space.

Mathematically this is achieved by representing possible solutions as coded strings. Many such strings are created, each representing a different location on the given search space. These strings are then evaluated according to some criteria, and the 'fittest' are given a higher probability of selection. Parts of the selected strings are combined to form new strings and occasionally part of the string is randomly assigned a new value. Eventually, just as animals adapt to their environment, the strings evolve to better match the criteria given. The method is similar to human search where good solutions receive more of our attention while bad solutions are less favoured. We would reasonably expect that combining and modifying parts of these existing good solutions may lead to better solutions and in some cases an improvement indeed on the original. An example of a simple GA is used in section 3.2 to demonstrate how the operators work together to provide the genetic search. First though, the differences between GA and classical search methods are discussed here.

Robustness is a central theme for all search algorithms. A balance between exploration of the search space and exploitation of available information is required in order to allow search algorithms to be successfully applied to a range of different problems. Traditionally many search methods have generally been calculus based, enumerative or random. Calculus based search methods work by finding points of zero slope. Generally this is achieved by stepping on the function and moving in a direction given by the steepest gradient. These methods are good for finding local optima which depend on the selected starting position. Furthermore, as the methods require gradient information, they are only applicable to functions with well defined slope values. This is a major drawback as many real life problems contain discontinuities and constraints which cannot be handled by these methods. Enumerative methods involve checking the function value at every point within the search space in order to find the optimal result. Such schemes are ideal for small search spaces but are highly inefficient for systems involving large search spaces or many parameters. Consider for example a case where we wish to identify N parameters, where each parameter has a search space consisting of 100 points. The total search space is then 100^N points and it quickly becomes impossible to evaluate the function at every point in a reasonable time. Even if we could evaluate a million points per second, the largest number of parameters that could be identified in this way within a year would be only 6, while it would take more than 3 million years to try all possibilities for 10 unknowns. Random search algorithms received attention as researchers recognised the shortcomings of calculus based and enumerative schemes. Nevertheless, they too are inefficient and in the long run can be expected to perform no better than an enumerative scheme with a coarser grid.

Genetic algorithms differ from the above-mentioned search methods in four significant ways.

(1) GA works with a coding of the parameter set rather than the parameters themselves. This is usually done using a binary system though other coding systems may also be used. This coding allows the GA to work in a very general way, allowing application to a wide range of problems. The coding does, however, present some problems. When binary coding is used, GA may find it difficult to move, or 'jump', between some values. These difficult jumps, known as hamming cliffs, may be observed by considering an example of a binary string of length 5 which may represent values form 0 to 31. The string 01111 would represent the

value of 15. If, however, the optimum is at a value of 16, the string required is 10000, a very difficult jump to make as all bits must be simultaneously altered. Alternative coding methods such as real number encoding used later in this book help to alleviate this problem.

(2)   GA search is carried out with a population of points, not a single point. Most optimisation techniques search from a single point, proceeding to the next point according to some predefined rule. These methods often fall on local optima and fail to find the desired global solution. GA searches using a population of many diverse points and as such is more likely to discover the global optimal solution.

(3)   GA uses an objective function rather than derivatives or other auxiliary information. Many other search techniques, particularly calculus based methods, require much information such as derivatives in order to work. GAs are "blind", only requiring the objective function (fitness values) in order direct the search.

(4)   GA works based on probabilistic rules rather than deterministic ones. Probabilistic rules are introduced to make the transition from one set of points to the next. This does not imply that GA is simply a random search, but means that GA uses random choice as a tool within a framework biased towards areas of likely improvement using information derived from the previous search.

A good summary of early GA works, and further details on how they differ from traditional search algorithms can be found in the very good book by Goldberg (1989). The combination of coding, a population of points, blindness to auxiliary information and randomised operators give GA the robustness required to solve a wide range of problems. It is noted here, however, that GA should not be treated simply as a black box, lest the computational time will become too large for solving realistic problems. Much understanding and refinements are needed to make the GA approach work effectively. Incorporating appropriate coding, altering the architecture of the GA, and integrating problem-specific information are essential in developing strategies appropriate to real world situations.

For illustration, a simple GA and its theoretical framework based on classical binary encoding and operators are presented in the following sections. Some of the modifications that have been made to improve the performance of GA are discussed in section 2.4, whereas the GA strategy developed and later applied in this book is described in chapter 3. Many have argued that new methods such as the ones presented in this book deviate from the original GA and as such use names such as evolution programs in order to acknowledge the deviation from traditional GA architecture, coding and operators. In this book, however, the term GA is still used. The reason is that, although the coding and architecture may not exactly resemble the original GA, the underlying principle remains the same.

2.2 A Simple GA

The concept of GAs is best explained by way of an example. This section uses the maximisation of a mathematical function to illustrate how a simple GA can be applied to search for the global optimal solution. Further examples and explanation can be found in Goldberg (1989) and Michalewicz (1994). Consider the problem of maximizing the function $f(x)$ as given in equation 2.1 over the range of $-20.0 \leq x \leq 20.0$.

This function, shown in figure 2.1, contains a global maxima at $x = 0$ and would be difficult to solve using classical optimization methods due to the many local maxima near the global optimal solution.

$$f(x) = 0.5 - \frac{\sin^2(2x) - 0.5}{1 + 0.02x^2} \tag{2.1}$$

The layout of a simple GA that may be used to maximize this function is shown in figure 2.2. A sample computer coding for this simple GA is included in the appendix to help the reader understand how the GA may be implemented. Parts of the code are also included in this section where necessary to demonstrate how the various operators are implemented. While the sample code is provided in FORTRAN, any language may be used and readers new to the area of GA are encouraged to write their own codes in a computer language they are comfortable with. It is also important to experiment with GA parameters in order to understand how the GA works and to observe the effect parameters may have on the performance of the GA.

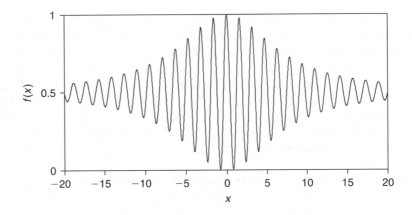

Figure 2.1 Function f(x) to be maximised

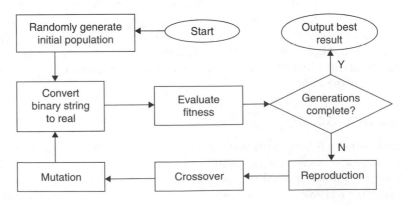

Figure 2.2 Layout of a simple GA

In this example, binary encoding is used. As the search range is $-20.0 \leq x \leq 20.0$, and in order to consider values to an accuracy of two decimal places, a binary string of length 12 is required. This binary number can represent integers from 0 to $2^{12} - 1$ and the binary to real conversion is made as shown in equation 2.2, where I is the integer represented by the N binary digits. LL and UL are the lower and upper bounds of the search space. For example the binary string 011001010001, represents the integer $I = 2^{10} + 2^9 + 2^6 + 2^4 + 2^0 = 1617$, and is converted to the real number $x = -4.21$.

$$x = LL + (UL - LL)\frac{I}{2^N - 1} \tag{2.2}$$

The GA begins by generating a set of initial candidate solutions. This is done by randomly assigning either a 0 or a 1 to each bit of each individual within the initial population. This is easily achieved using a random number generator as shown in the extracted code below. The population here contains Pop_size number of individuals of length N. All bits in the population are initially assigned a 0 value. For each bit, a random number r in the range [0 1] is then generated and if the value is greater than 0.5, the bit is changed to a value of 1.

Random generation for initial population

```
Pop=0
DO i=1, Pop_size
    DO j=1, N
        CALL RANDOM_NUMBER(r)
        IF (r>0.5) Pop(i,j)=1
    END DO
END DO
```

The binary strings are converted to real numbers using equation 2.2 and then the fitness of each solution is calculated. The fitness, or objective function, is a measure of the quality of a given individual. As the objective in this case is to maximize $f(x)$, and $f(x)$ is greater than 0 for all values of x, the function value gives an indication of the quality of the solution and can be used directly as the fitness function.

Reproduction, or selection, is designed to select fitter individuals to receive greater representation in future generations. Many different selection procedures, including both probabilistic and deterministic sampling, may be used. One simple way to carry out reproduction is the so called roulette wheel method, shown in the code below. Each individual is assigned a selection probability proportional to its fitness and cumulative probabilities are computed for the selection phase. The new population is then selected by 'spinning the wheel' the required number of times. Each time the wheel is spun, an individual is selected by comparing the random number with the cumulative probabilities. In this way selection is made with replacement until the new population (T_pop) is full. This method encourages multiple selections of fitter individuals and filters out the weakest individuals.

Selection by roulette wheel method

```
P_select=Fitness/SUM(Fitness)
DO i=2,Pop_size
    P_select(i)=P_select(i)+P_select(i-1)
END DO

DO i=1,Pop_size
    CALL RANDOM_NUMBER(r)
    DO j=1,Pop_size
        IF (P_select(j)>=r) THEN
            T_Pop(i,1:N)=Pop(j,1:N)
            EXIT
        END IF
    END DO
END DO
Pop=T_Pop
```

Crossover and mutation allow the GA to discover new solutions. In this example, a simple crossover is used. The crossover rate determines the chance of an individual being involved in a crossover and, once selected, two individuals (parents) are paired up for the crossover to take place. The crossover point is randomly selected and the ends of the parents are switched to form two new individuals (offspring). For example, if the parent strings 111000111001 and **100011100001**, representing the values $x = 15.57$ and $x = 2.20$, are crossed after the 4th bit, the offspring created are 111011100001 and **100000111001**, representing the values $x = 17.21$ and $x = 0.56$. There are several ways to select and pair the parents for crossover. For the method shown in the code below, parents are first selected into a crossover pool according to the given crossover rate. Once selected, the order of the parents is shuffled in order to randomly assign partners. The shuffle subroutine can be seen in the full code provided in the appendix. If the number of individuals selected is odd, one of the parents is discarded. Crossover is then carried out using the selected pairs using a random crossover point and the offspring replace the parents in the population.

Crossover

```
j=0
DO i=1,Pop_size
    CALL RANDOM_NUMBER(r)
    IF (r<P_cross) THEN
        j=j+1
        Select(j)=i
    END IF
END DO
CALL Shuffle(Select(1:j),j)
```

```
IF (MOD(j,2)==1) j=j-1

DO i=1,j,2
    CALL RANDOM_NUMBER(r)
    cross=CEILING(r*(N-1))
    O1(1:cross)=Pop(Select(i),1:cross)
    O2(1:cross)=Pop(Select(i+1),1:cross)
    O1(cross+1:N)=Pop(Select(i+1),cross+1:N)
    O2(cross+1:N)=Pop(Select(i),cross+1:N)
    Pop(Select(i),1:N)=O1
    Pop(Select(i+1),1:N)=O2
END DO
```

The crossover operator simply recombines information which already exists, but is unable to explore areas not included in the population. For example, the parents above both contain a 0 at position four and no crossover can change this value to a 1. Mutation is therefore needed to ensure the whole search space can be explored. Mutation works by changing individual bits from 1 to 0 or vice versa. The chance of an individual bit being mutated is determined by the mutation rate and all bits are treated in the same way. For example if the second and seventh bits of the individual 111000111001 undergo mutation it will become 101000111101.

The whole process of fitness evaluation, reproduction, crossover and mutation is repeated for a given number of cycles, or 'generations', and the best solution obtained is output. As an illustration, the simple GA described above is applied using a population of 10, crossover rate of 0.8, mutation rate of 0.05 and 50 generations. The best solution at the end of each generation is recorded and plotted in figure 2.3 to illustrate how the GA evolves the solution over time. In the figure it is seen that the solution quickly converges to a local maxima of 0.976 at $x = 1.56$ which is on the local maxima nearest the global solution. It is also observed that the solution is able to 'escape' the local

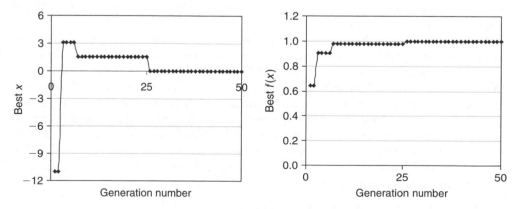

Figure 2.3 Function maximisation – GA solution

maxima to a value of $x = 0.024$ in the 26th generation, before improving to $x = 0.005$ as the final result.

This example highlights an important feature of GA. A major strength of GA is the better capability to escape from local optima to find the global optima solution compared to many other methods. Nevertheless, while the global maximum solution is found in this case, this may not always be the case. The identification above is repeated a total of 50 times. Of those, a solution on the global peak is discovered 32 times, while the first and second local peaks are discovered 11 and 7 times respectively. In developing a GA, the reliability and robustness of the solution is therefore very important owing to the stochastic nature of the search process. Of course we can increase the population size and the number of generations, but at the cost of longer computational time. It is also possible to influence the search by selecting appropriate crossover and mutation rates, but in general there is a trade off between exploration (broad search) and exploitation (local search). For example, small crossover and mutation rates will help explore the domains around the current solutions and will be less likely to destroy good solutions. It will, however, make it harder to explore new domains. Large crossover and mutation rates, on the other hand, will help cover more ground, but at the expense that the good solutions will be less likely to develop further and will find it harder to converge. This trade off between exploration of the search space and exploitation of promising solutions has long been an issue with simple GAs and is one of the key motivations behind the improved strategy presented in the following chapter.

2.3 Theoretical Framework

The simple GA used in the previous example, adopts binary encoding of variables and simple crossover and mutation operators. Early attempts to explain why GA worked used the idea of schema as the building blocks of the solution. This theory is able to show how favourable building blocks can survive and prosper in a GA and hence how a population could improve over time. This classical theory has received considerable criticism (e.g. Koehler, 1997) as it does not consider how a GA is able to search outside the information present within the population. In addition the theory is too simplified to explain the complex operators and real encoding used in the algorithms developed and applied in the later chapters of this book. It is nevertheless the basis of why GA works and as such a basic summary of the theory is included here. More detailed discussions on schema and GA theory can be found in Goldberg (1989).

A schema is created by introducing a 'don't care' symbol (*) into the alphabet to indicate positions which could be filled by any value. For example in a binary chromosome of length 10 the schema $(1**0******)$ would represent all individuals with a 1 in the first position and a 0 in the fourth. Schema vary depending on which positions are fixed (0 or 1) and which are free (*), and as such it is useful for us to have a way of defining certain properties of the schema. The number of fixed positions (0s or 1s) gives the *order* of schema S, $o(S)$ which tell us how well defined a schema is. A high order schema is therefore more specific about the group of strings it describes. The distance from the first to last fixed position is the *defining length* $\delta(S)$. An example of these parameters is given for the strings S1 and S2 shown below.

$$S1 = (*1011*****) \quad o(S1) = 4 \quad \delta(S1) = 3$$
$$S2 = (1******00*) \quad o(S2) = 3 \quad \delta(S2) = 8$$

The theory is concerned with determining the number of a given schema present in subsequent generations. That is if the number of a given schema S present in the population at time t is denoted as $\xi(S, t)$, what will be the likely number of schema present at time $t + 1$, $\xi(S, t + 1)$? There are two factors of consideration, i.e. the selection of the schema, and the possible destruction of schema due to crossover and mutation operations.

For standard roulette wheel selection, the selection process is based on fitness, where the number of a given schema selected is proportional to the average fitness of the individuals represented by the schema $f(S, t)$, compared to the average fitness of all the individuals in the population $\overline{f(t)}$. That is, it would be expected that based on selection alone,

$$\xi(S, t + 1) = \xi(S, t) \cdot \frac{f(S, t)}{\overline{f(t)}} \tag{2.3}$$

This shows that, considering selection alone, it is expected that schema with above average fitness will receive an increasing representation in successive generations.

The crossover operation is designed to combine promising building blocks. However it can also destroy them. For example, consider the schema S1 and S2 and the individuals b1 and b2 below.

S1 = (*1011*****)   S2 = (1******00*)

b1 = (0101100111)   b2 = (1110101001)

String b1 is an example of schema S1 and b2 an example of S2. If these two individuals are selected for crossover and crossed after the fifth bit, the resulting offspring would be

b1' = (0101101001)   b2' = (1110100111)

It is clear that the schema S1 survived in b1' but schema S2 did not. Additionally b2' contains neither S1 nor S2. Some probing into this reveals that a schema can be destroyed only if the crossover point is within the range enclosed by the fixed bits of the schema. That means that the chance of a schema surviving the crossover is dependant on its defining length $\delta(S)$. In the above example we have $\delta(S1) = 3$ and $\delta(S2) = 8$. If the total length of the string is m, there are $m - 1$ possible crossover locations and therefore the chance of crossover within the schema is $\delta(S)/(m - 1)$. It is therefore reasonable that in the above example, S1 with a chance of destruction of 3/9 survived, whereas S2 which had an 8/9 chance was destroyed. It is of course possible that a crossover within the schema may not destroy it, or that a crossover may create an example of the schema where it previously did not exist. For this reason, the above is treated as a 'worst case' or lower bound to the expected behaviour. Additionally as individuals are selected for crossover with a probability $p_c \leq 1$, it may be that not all individuals will be subject to this possibility of destruction. The expected number of schema considering both selection and crossover is then updated as

$$\xi(S, t + 1) \geq \xi(S, t) \cdot \frac{f(S, t)}{f(t)} \cdot \left(1 - p_c \frac{\delta(S)}{m - 1}\right) \tag{2.4}$$

Note the inequality ($\geq$) is now used to account for the possible creation of new examples of the schema through crossover.

Finally the mutation operator is considered. Again the chance of destruction of the schema is calculated and the fact that schema may be created is absorbed by the inequality. During simple mutation each bit is mutated with a probability, p_m. Thus for each bit the chance of survival is $(1 - p_m)$. Each bit is subject to this same chance of mutation and hence for a schema of order o(S) the chance of survival of the schema is $(1 - p_m)^{o(S)}$. In general the mutation rate is low and this can be approximated as $1 - p_m \cdot o(S)$. The theory is then complete as,

$$\xi(S, t+1) \geq \xi(S, t) \cdot \frac{f(S, t)}{f(t)} \cdot \left(1 - p_c \frac{\delta(S)}{m-1}\right)(1 - p_m \cdot o(S)) \tag{2.5}$$

Furthermore, as the final term in the expansion of the above is small,

$$\xi(S, t+1) \geq \xi(S, t) \cdot \frac{f(S, t)}{\overline{f(t)}} \cdot \left(1 - p_c \frac{\delta(S)}{m-1} - p_m \cdot o(S)\right) \tag{2.6}$$

This result shows that highly fit, short, low order schemata will receive increasing representation at each generation. Further to this, the theory makes the assumption that the relative fitness of a given schema and the overall population remains constant. That is,

$$\frac{f(S, t)}{\overline{f(t)}} \cdot \left(1 - p_c \frac{\delta(S)}{m-1} - p_m \cdot o(S)\right) = 1 + \varepsilon \quad \text{for all t}$$

$$\tag{2.7}$$

$$\rightarrow \xi(S, t) \geq \xi(S, 0) \cdot (1 + \varepsilon)^t$$

Thus the highly fit, short, low order schemata receive *exponentially* increasing representation in the population.

2.4 Advances in GAs

Over the past three decades various forms of GAs have been widely developed and applied. A basic coding using binary representation and set of operators, mutation, crossover and reproduction formed the early basis for application into mathematical problems. Later as application moved into more complex areas new coding schemes and operators were developed to adapt to the problems under study. In recent years, much effort has also been made to alter the architecture of GAs and to incorporate local search algorithms to further improve the performance and to help reduce the problems associated with standard GA where a trade off exists between exploration and exploitation of the possible solutions.

The foundations for GAs were laid by Holland and his students in the early 1960s (Holland 1962a-b) with a mathematical framework and the idea of schema following shortly after (Holland 1968, 1971, 1973). By the time Holland collated his ideas in his 1975 book 'Adaptation in Natural and Artificial Systems', the basics of GAs were well established. Though Holland is unquestionably the father of GA, the first use of the term 'genetic algorithm' was in fact by one of his students (Bagley 1967). While Holland's work remained general, another one of his students (De Jong 1975) began to focus on problems in mathematical function optimization. De Jong reduced the genetic algorithm to its bare essentials in order to conduct an in-depth study into the effect of genetic operators. The resulting GA, using simple crossover and mutation and roulette wheel selection was denoted as R1. In addition, De Jong considered five additional models, R2 to R6, which used various modifications of the genetic operators. The study, on mathematical test functions, paved the way for future GA studies and applications. A very good review of the early development of genetic algorithms and a collection of some influential papers can be found in Goldberg (1989) and Fogel (1998) respectively.

Following De Jong's work, a number of studies were conducted on improving the basic GA. The crossover and selection operators were often the focus, with several procedures proposed with regards to crossover (Booker 1987), selection (Baker 1987), fitness scaling (Goldberg 1989), and ranking (Whitley 1989). These modifications attempted to improve performance by striking an appropriate balance between exploration and exploitation of solutions. For example, using fitness scaling techniques, diversity can be maintained in the population during early stages by reducing the impact of highly fit individuals, while late in the process when fitness values tended to converge, differences in fitness can be exaggerated to ensure higher success of better individuals in the probabilistic selection.

Due to their general form, GAs have been applied to a wide range of problems. *Mathematical function optimization* problems have generally been used in the development of GAs due to ease of implementation and direct calculation of fitness. The five-function test suite of De Jong (1975) has often been used, and was extended to ten functions by Schaffer et al. (1989). Their study, on the effect of GA parameters, suggested that mutation may play a more crucial role than had previously been recognized. The F6 function proposed by Schaffer was used to demonstrate a modified GA proposed by Potts et al. (1994). The modified GA split the population into 'species' allowing for different rates of mutation and crossover to be used in each species. This division of the search allowed broad exploration of the search space to be conducted in parallel with a search exploiting the best solutions. The improved algorithm proposed in the next chapter uses this idea of multiple species, while modifying the strategy to use real encoding and improved genetic operators. *Combinatorial optimization* using GAs often focused on the travelling salesman problem (TSP). The TSP is conceptually a very simple problem whereby a salesman wishes to visit n cities and return home in the most efficient sequence. As the number of cities increases, this problem quickly becomes difficult due to the large search space as the number of possible combinations is given by $n!$. A good overview of the use of GAs for TSP is given in Michalewicz (1994) while some of the early efforts in this area can be seen in Goldberg and Lingle (1985), and Grefenstette et al. (1985). *Game theory* problems such as the iterated prisoner's dilemma are well handled by GA as various strategies develop and compete for survival (Axelrod 1987).

System identification problems are solved using GAs by specifying an appropriate objective function, usually specified in a form that rewards smaller errors between simulated and measured system output. The identification of linear and non-linear auto regressive with exogenous inputs systems using GAs has been studied by Luh and Wu (1999). Iba et al. (1993) presented results for non-linear time series prediction and pattern recognition problems which used a GA combined with a least squares method. The GA was used to develop an appropriate model while the least squares method was used to find appropriate model coefficients.

The use of GAs in *structural identification* and *damage detection* is a relatively new development. Much of the work on structural identification has been carried out by the first author and his colleagues and students, incorporating GAs into various substructure and hybrid identification schemes (Koh et al. 2000, 2003a,b, Koh and Shankar 2003a,b). These schemes generally aim to identify stiffness and damping (and mass) parameters from the dynamic time history information with an objective function that minimizes the error between the measured and simulated accelerations. Some success has also been achieved using static displacements or frequency domain models. Perera and Torres (2005) identified damage in structures by minimising a dynamic residue vector, while Koh and Shankar (2003a,b) used reference displacements from a frequency based dynamic model. Rao et al. (2004) also used frequency information, utilizing the sum of diagonal terms from a residual force matrix as the objective function. Chou and Ghaboussi (2001) simply used the response of the structure to a series of static load to define their objective function. This method nevertheless has a limitation that only stiffness information can be obtained. The evolution strategy proposed by Franco et al. (2004) was in effect an adaptive GA, whereby the magnitude of mutations adapted as the analysis proceeded. The results presented for a 10-DOF structure were very good. Where full output was available the average error was only 2.7% under 5% noise. The procedure failed, however, when only partial output of three measurements was used and the average error increased to more than 15%. The modified GA strategy presented and applied throughout this book was first proposed by the authors for structural identification problems (Perry et al, 2006). Using a combination of a search space reduction method and a novel modified GA, the strategy is able to accurately and efficiently identify parameters. This strategy is discussed further in chapter 3, and applied to a variety of problems in the subsequent chapters.

A note is in order here on the role of local search in GA. Research works have shown that local search is a useful tool to complement the GA in improving the fine tuning capability. The accuracy and robustness can be greatly enhanced by embedding GA-compatible local searchers (Koh et al. 2003b). But the local search is usually executed in the inner loop of a general GA; thus the accuracy is generally achieved at the cost of computational time. Recently a Levenberg-Marquardt (LM) method of local search was proposed (Kishore Kumar et al. 2007) and found to give good performance in identifying a 3-DOF nonlinear system. Nevertheless, this local search method appears to be not suitable for large system identification because LM has to store the approximate Hessian matrix which can be large and expensive in its repetitive inversion. Besides, considerable preliminary GA runs are needed to provide sufficiently good initial guess, particularly for systems with large number of unknowns. Hence, the approach of embedding a local searcher is not adopted here. Instead, the multi-species method to be presented in the next chapter facilitates local search in a seamless way by controlling the mutation rate.

2.5   Chapter Summary

Genetic algorithms have been introduced in this chapter as the search tool for the optimal solution. Using a structured yet random search, this method has been shown to possess several crucial advantages over classical methods for structural identification.

GAs differ from traditional search algorithms in several ways: (1) they work with a coding of the parameter set rather than the parameters themselves, (2) they search from a population of points, not a single point, (3) they use an objective function (which can be defined in terms of any response quantify and gives flexibility) rather than derivatives or other auxiliary information, (4) they work based on probabilistic rules rather than deterministic ones. Moreover, GAs have a high level of concurrency and is thus suitable for distributed computing.

The combination of coding, a population of points, blindness to auxiliary information and randomised operators give GAs the robustness required for application to a wide range of problems. Nevertheless GAs should not and cannot be treated as a black box; lest the computational time would be too prohibitive for real problems. Much understanding and refinements are needed to make the GA approach work effectively. While some issues such as the trade off between exploration and exploitation of solutions still remain, recent application of GAs to system identification problems has shown some promising success. It is the purpose of this book to build on recent advances and to demonstrate how robust and efficient strategies may be developed and applied to identify parameters and detect damage in engineering structures.

Chapter 3

An Improved GA Strategy

The identification strategy presented in this chapter is in fact an iterative, two-tier strategy as illustrated in figure 3.1. At the fundamental tier, an improved GA based on migration and artificial selection (iGAMAS in short) is used to identify the system based on a given set of search space limits. At the higher tier, a search space reduction method (SSRM) makes use of the results from iGAMAS to reduce the search space, feeding the new limits back to the iGAMAS for use in the next identification cycle.

The motivation behind the development of the SSRM arises from the fact that for GAs, the convergence rate and accuracy are highly dependent on the size of the search space. By adaptively reducing the limits of the search, a more accurate and efficient identification is possible. The heart of the method is the iGAMAS. This algorithm, based on the GAMAS proposed by Potts et al (1994), has been developed in order to provide a good identification technique that simultaneously explores the search space and focuses on promising individuals. The proposed iGAMAS includes a reduced data length procedure and other novel features that are introduced to greatly reduce the computational time and to increase the accuracy of identified parameters. The SSRM strategy presented in this chapter is designed to be applicable to a wide range of problems, be it financial, mathematical, biological, structural, hydrodynamic etc. As long as the system can be represented by a reasonably accurate model, capable of reproducing the systems response to a given input, the SSRM can be used. This broad applicability is an advantageous feature of GA based algorithms and is displayed in the later chapters, where the SSRM is applied to various problems including output-only

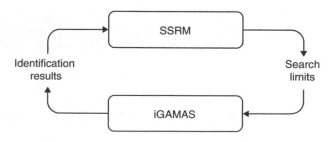

Figure 3.1 Overview of GA Strategy

identification and substructural identification by making appropriate adjustments to the model and formulation.

3.1 SSRM

The search space reduction method (SSRM) is designed specifically to increase the accuracy and efficiency of identification by reducing the search space. The layout of the SSRM is shown in figure 3.2, while the iGAMAS, that provides the search capability, is shown in figure 3.5 and explained in the next section. The basic idea behind the SSRM is as follows; adaptively reduce the search space for those parameters that converge quickly in order to reduce the computational effort spent looking far outside the area where the optimal solution lies. This is achieved by carrying out several runs of the iGAMAS, following which the mean and standard deviation of the identified parameters are computed. The standard deviation gives us an indication of the uncertainty of the identified parameter and the search space can be reduced accordingly. If the standard deviation is small it is likely that the mean is close to the optimal parameter value and the search limits can be reduced. Conversely, if the standard deviation is large we should continue to search broadly for that parameter.

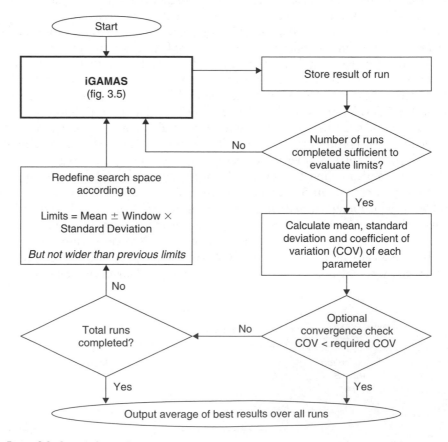

Figure 3.2 Search Space Reduction Method

Eventually as some parameters converge almost exactly, the SSRM in effect reduces the number of unknown parameters and those remaining can be identified more efficiently.

The main parameters that define the SSRM are the number of *runs* to be used for evaluation of the search space, the *total runs* to be carried out and the width of the reduced search space *window*. In addition a *convergence* exit criteria may be included to exit from the system early if satisfactory convergence is achieved. The final result can be output as either the single best result or as the average of a given number of the best runs over all of the runs conducted. Our experience based on many numerical studies shows that, for the structural identification problems studied in this book, there is no significant benefit in keeping additional results. In these cases, using the single best result as output gives the best performance.

3.1.1 Runs for Evaluation of Limits and Total Runs

In order to be effective, the SSRM must work on a reasonable number of runs. The number of runs to be used for evaluation of the search space limits should consider the following points.

- The number of runs must be sufficient to get a reasonable estimate for the mean parameter value.
- The search space is not reduced until the given number of runs is complete. A large number of runs therefore delays the time when the search space is first reduced.
- Newer results should be more accurate but including a large number of results in the evaluation of new limits may actually impede convergence. On the other hand, too few results may cause premature convergence to local optima.

The number of runs must therefore be carefully selected to achieve the desired performance as more runs will make the system more robust, but will slow convergence resulting in less accurate results and/or an increase in the total computational time.

The total runs to be used depends mostly on the accuracy required and the computational time allowed. In theory, as the search space is reduced after each additional run, the results will become more and more accurate. In reality, however, accuracy will be limited due to such factors as noise, and after a time no further improvement in accuracy is possible. The total runs should also consider other factors such as the population size and number of generations. For example, if all other parameters are constant, using a total of 10 runs and 200 generations per run will result in the same computational time as would 20 runs with 100 generations per run. It is therefore important to achieve a good balance of GA parameters. Suggested GA parameter values are discussed in Chapter 4.

3.1.2 Reducing the Search Space

Search limits = Mean ± window × std dev; but not beyond original limits (3.1)

The width of window defines how quickly the search space is reduced according to equation 3.1. It is important to choose a window that is small enough to encourage convergence but big enough so that the global solution is very likely to lie within the new, reduced, search space. Again this window parameter will depend on other

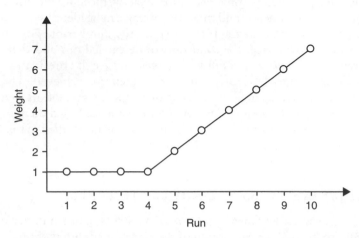

Figure 3.3 Example of weights used if 4 runs are used to evaluate search space and 10 total runs are allowed

GA parameters as well as the nature of the problem. For simple problems, where the results are expected to be quite reliable, a smaller window can be used, whereas if the results are uncertain a larger window may be a safer option. In the application of the strategy in this book the window is taken as 4, i.e. the reduced limits are set as 4 standard deviations on each side of the mean. Statistically this ensures there is a very high chance that the actual result will remain within the reduced limits.

In the SSRM the mean value of each parameter is calculated using weighted results whereby the more recent runs are given a higher weighting. This is to recognise that the results should improve as the search space is reduced. The weighting used is as follows. A weight of 1 is assigned to each of the original runs, until the first time the search space is reduced. The run immediately following this is given a weighting of 2, then 3, 4 and so on as illustrated in figure 3.3. For example, if 4 runs are used for the evaluation of limits and a total of 10 runs are used, the search space will first be evaluated after the 4th run. In this case all four runs will be weighted equally. After the fifth run, the limits will again be evaluated, with the fifth run assigned a weighting of 2 and the second, third and fourth runs each assigned a weight of 1. By the time the identification reaches the end of the ninth run, the limits used for the final (10th) run will be evaluated from runs six to nine with weights of 3, 4, 5 and 6 respectively.

If desired, a convergence criterion can be included so the results can be output early if the values converge quickly. In this case the ratio of the standard deviation and mean (coefficient of variation) is used. In general the coefficient of variation gives an indication of the potential error in the parameter values and is therefore useful to check at the end of the program to see to what extent the results have converged.

3.1.3 *Example of Function Maximisation*

$$f(x_1, x_2) = 100 - 100x_1^2 + 1000x_1 - x_2^2 + 2x_2 \tag{3.2}$$

As an illustrative example of the SSRM procedure consider maximisation of the two-variable function given in equation 3.2, over the range $0 \leq x_1 \leq 10$, $0 \leq x_2 \leq 10$.

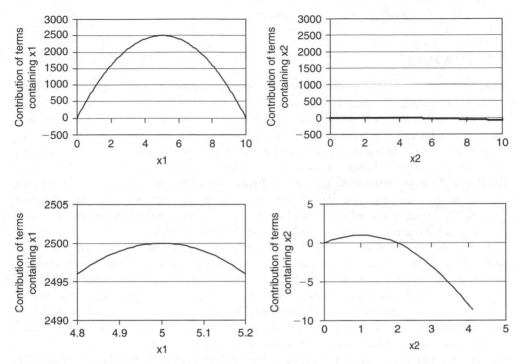

Figure 3.4 Variation of function due to x_1 and x_2 Original search space above, reduced search space below

The function has a global maximum value of 2601 at (5,1). The challenge in this problem is that the variables have largely different effects on the function value. Over the range given, the terms containing x_1 contribute values ranging from 0 to 2500, whereas the terms containing x_2 contribute values ranging from -80 to 1. This results in a common problem faced by many search strategies whereby changes in x_1 alter the function value far more than changes in x_2. Consequently, the parameter value identified for x_1 will generally be much more accurate than that for x_2. The SSRM tackles this problem very effectively. For example, the first four runs of the iGAMAS may return results of (4.96, 0.20), (4.99, 1.80), (5.00, 1.70) and (5.08, 0.92), with corresponding function values of 2600.20, 2600.35, 2600.51 and 2600.35. These results give mean parameter values of 5.0075 and 1.155 and standard deviations of 0.0512 and 0.7484.

It can been seen that, while x_1 has converged almost exactly, x_2 has considerable variation. The SSRM would then reduce the limits. Using a window width of 4.0 the new limits become $4.8027 \leq x_1 \leq 5.2123$, $0 \leq x_2 \leq 4.1486$. Within these new limits the function value varies by up to 4.51 due to changes in x_1 and by 9.91 due to changes in x_2. The relative significance of the parameter x_2 has increased and the identification results in future runs will improve accordingly. The parameter x_1 will also continue to improve as the search becomes more focussed and effort is not wasted evaluating values that lie far from the optimal solution. The reduction in search space and variation in function value is shown graphically in figure 3.4. In the plots, the vertical scale has a range of 3500 in both of the upper plots and 15 in the lower plots so valid comparison of the significance of x_1 and x_2 can be made.

3.2   iGAMAS

The heart of the SSRM is the improved genetic algorithm utilising migration and artificial selection (iGAMAS). This strategy is based on the basic GAMAS by Potts et al (1994) but uses a floating-point representation and includes new operators and techniques designed to increase the speed and accuracy of identification. The basic layout of the iGAMAS is shown in figure 3.5 and the important features of the strategy are discussed in the subsequent sections. The most important features that distinguish the IGAMAS from 'normal' GA are the inclusion of multiple species, artificial selection, regeneration and a variable data length procedure. In addition to these important points, the strategy includes a rank based selection, new mutation operators and a new tagging procedure to ensure diversity in the best solutions.

The basic layout of the GAMAS and iGAMAS algorithms are similar, with both utilising multiple species, and an artificial selection procedure. The major difference comes in the way the search is conducted within each species. The original GAMAS is a binary coded GA using classical crossover and mutation operations. The search is controlled by allowing for a different rate of mutation in each species. In contrast, the iGAMAS proposed here uses real encoding of variables and as such adopts non-uniform mutation operators, allowing the focus of the search to vary, not just across species, but also over time. In addition to this major difference, the iGAMAS includes a new tagging procedure and a reduced data length procedure which is specifically designed for dynamic problems. These important changes and additions are discussed in the relevant sections below.

3.2.1   Solution Representation

Solutions are represented using floating-point numbers in vector form. Each parameter is represented by a single value and the vector of all parameters makes up an individual (possible solution). The floating-point representation is more natural and compact than the binary encoding traditionally used in GA, and avoids problems such as hamming cliffs discussed in the previous chapter. In addition, the floating-point representation enables easy application of new operators, such as non-uniform mutation, that would have been more difficult or impossible to implement in a binary system. There are of course some arguments in favour of binary encoding, the main one being that, by controlling the number of bits, we can effectively control the resolution of the search required. This is however considered a minor benefit, or even a disadvantage, with respect to the problems under study in this book. This is because a reasonably high resolution is required to properly capture the dynamic behaviour of the system.

In the computer programs developed, double precision (8-byte) real numbers are used allowing for approximately 15 significant digits. Each species is stored in matrix form with each row in the matrix representing an individual solution. Figure 3.6 gives an illustration of the representation for a problem of m variables, x_1 to x_m, and species size n.

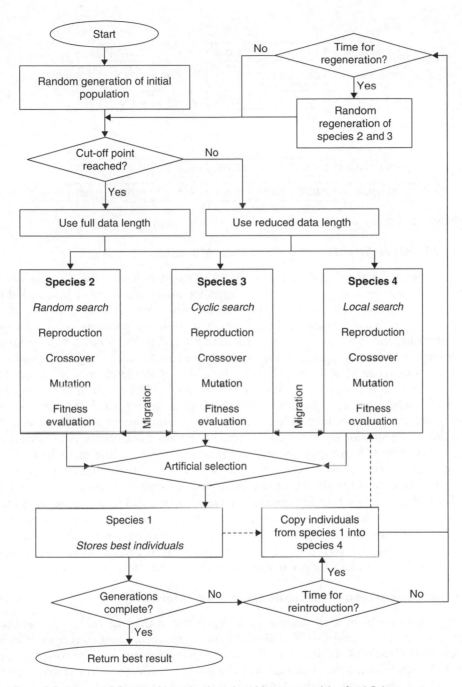

Figure 3.5 Improved Genetic Algorithm based on Migration and Artificial Selection

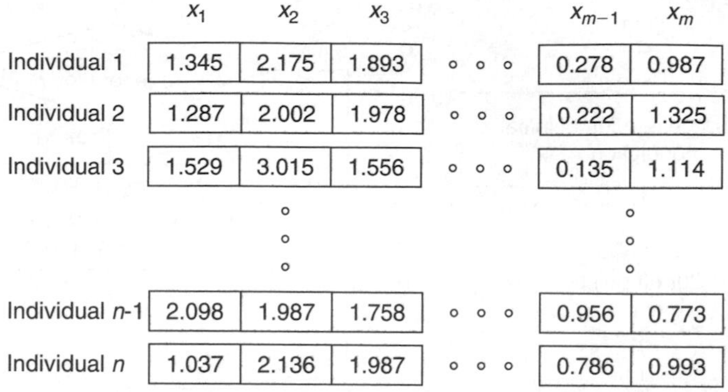

	x_1	x_2	x_3		x_{m-1}	x_m
Individual 1	1.345	2.175	1.893	o o o	0.278	0.987
Individual 2	1.287	2.002	1.978	o o o	0.222	1.325
Individual 3	1.529	3.015	1.556	o o o	0.135	1.114
		o			o	
		o			o	
		o			o	
Individual n-1	2.098	1.987	1.758	o o o	0.956	0.773
Individual n	1.037	2.136	1.987	o o o	0.786	0.993

Figure 3.6 Example of representation and storage of solutions

3.2.2   Multiple Species and Focus on Mutation

The real power of the GAMAS and iGAMAS strategies lies in the division of the population into species. One of the challenges for many search strategies, including GAs, is the trade off between exploration and exploitation. That is, it is difficult to find a balance between utilising the information from the previous good solutions (exploitation), and maintaining a broad search capability (exploration). By splitting the population into multiple species, this problem can be adequately addressed. As one species searches broadly, another can be designed to search locally around the best solutions. In the iGAMAS, four species are used. Species 1 is used to store the best results while species 2–4 conduct searches increasing in focus from a very broad random search to a more refined local search. The various searches are controlled by using different mutation operators. The focus on mutation is a necessary modification from the original GA brought about because of the real encoding of variables. In a binary system, crossing over creates new parameter values and mutation is necessary mainly to ensure that specific bit values are not permanently lost from the process. In a real coded system however, crossover effectively becomes, a recombination operator, only altering the combination of parameters and not the parameter values themselves. Mutation therefore becomes highly important in modifying the existing parameter values in order for the search space to be properly explored. The operators used to achieve the desired mutations are discussed further in the relevant sections.

3.2.3   Regeneration, Reintroduction and Migration

A well-known problem with many search algorithms is that the solutions may converge to local optima and find it difficult to escape from the local convergence basin in order to find the global optimum solution. Regeneration involves the complete random replacement of a species. In this way the process is effectively restarted and new optimum may be found. In the iGAMAS strategy developed, only species 2 and 3 are regenerated. This allows species 4 to focus on refining the previously generated solutions while species 2 and 3 search for new possibilities. The number of times regeneration is carried out must be reasonably small to allow sufficient time between regenerations for good solutions to develop.

To ensure that species 4 operates on a set of good solutions, a reintroduction is required. This involves inserting individuals from species 1 into species 4 at a prescribed interval. The number of reintroductions required should consider that, while it is desirable to have the best results being modified in species 4, some time may be needed in order to develop the solutions. Despite this, it is found that ensuring the best results are present in species 4 is very important and reintroduction should be carried out frequently to achieve the best results.

Migration facilitates the exchange of information between species. Just as human movements between different countries (or companies) can help transfer knowledge and ideas, the migration of individuals between species can help share important information. The migration operation involves swapping randomly selected individuals between species 2 and 3 and also between species 3 and 4. The number of individuals to be moved at each generation is controlled by the migration rate, which in this book is taken as 0.05, meaning 5% of the species is exchanged at each generation.

3.2.4 Mutation Operators

A key benefit of multiple species and floating-point representation is that different mutation operators can be employed to compliment one another. Three different mutation operators are used for species 2 to 4 in the iGAMAS. The mutation operators are designed to give each species a different strength so that the whole system can be effective. In each case mutation is carried out on a single parameter value. The mutation rate determines the probability of an individual value being mutated and a random number generator then determines the magnitude of the mutation to be applied. A graphical representation of the mutation provided by the operators for species 3 and 4 is shown in figure 3.7 for a case where regeneration is carried out 3 times and the random number generated is 0.5. Note that the completely random mutation of species 2 would have a value of 0.25 in this case and would be independent of the generation number.

3.2.4.1 Species 2 – Random Mutation

Species 2 is designed to search broadly to uncover promising areas in the search space that have not yet been discovered. The random mutation of species 2 simply involves random regeneration of the value. The selected parameter within the individual is assigned a value randomly distributed within the parameter limits by generation of a random number, r in the range [0 1] as

$$x_i = LL_i + r \times (UL_i - LL_i) \tag{3.3}$$

where UL_i and LL_i are the upper and lower limits of the search space for the ith parameter x_i.

3.2.4.2 Species 3 – Cyclic Non-Uniform Mutation

The non-uniform mutation operator reduces the average magnitude of mutations as the analysis proceeds and has been shown to help increase the accuracy and convergence rate in mathematical optimisation problems (Michalewicz 1994).

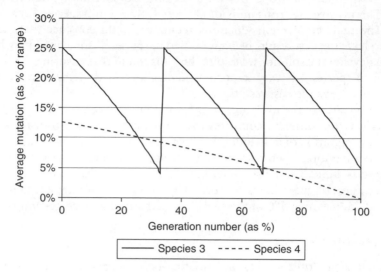

Figure 3.7 Average magnitude of mutations for species 3 and 4

The *cyclic* non-uniform mutation operator proposed here is based on this operator but is modified with the regeneration procedure in mind. The idea is to allow for larger mutations after regeneration has taken place and then to gradually reduce the size of the mutations as the solutions develop. This means the average size of the mutations will decrease gradually within each regeneration cycle to allow solutions to improve around the current values, and then increase again after the regeneration, where it is desirable to search broadly again for new possibilities. To achieve this objective the following operator is used:

$$x_i = x_i + (UL_i - x_i) \times \left(1 - r_1^{\left(1 - \frac{0.9 MOD(g,R)}{R}\right)}\right) \quad \text{for } r_2 = 0$$

$$= x_i + (LL_i - x_i) \times \left(1 - r_1^{\left(1 - \frac{0.9 MOD(g,R)}{R}\right)}\right) \quad \text{for } r_2 = 1 \tag{3.4}$$

where r_1 is a random number in the range [0 1] and r_2 is randomly selected as either 0 or 1. $MOD(g,R)$ is the remainder when the generation number g is divided by the number of generations between regenerations, R. As species 4 is reserved for local search, the factor of 0.9 ensures that the average size of the mutations in species 3 is not too small as the generations approach a regeneration point.

3.2.4.3 Species 4 – Local Non-Uniform Mutation

As species 4 is designed to refine the best solutions, small mutations are preferred. A local non-uniform mutation operator is used, whereby the size of mutations is gradually reduced as the analysis proceeds. The following non-uniform operator achieves this mutation.

$$x_i = x_i + 0.5 \times (UL_i - x_i) \times (1 - r_1^{(1-g/G)}) \quad for \ r_2 = 0$$
$$= x_i + 0.5 \times (LL_i - x_i) \times (1 - r_1^{(1-g/G)}) \quad for \ r_2 = 1 \tag{3.5}$$

where G is the total number of generations to be run and the multiplier (0.5) ensures smaller mutations as illustrated in figure 3.7.

3.2.5 Crossover Operators

Two types of crossover are used in the iGAMAS strategy developed, namely a simple crossover and a multipoint crossover. The crossover operators used do not alter the values of individual parameters and should be thought of as 'recombination' operators as they recombine parameters from different individuals. The mutation and crossover operators therefore work in tandem, modifying and recombining the parameters to explore new areas of the search space.

3.2.5.1 Simple Crossover

The simple crossover is similar to the crossover performed in binary GAs, the only difference being that, as the parameters are represented by real numbers, crossovers can only occur between parameters and as such cannot alter the parameter values themselves. Where mutation is carried out on each parameter separately, crossover is applied to whole individuals. The probability of an individual being involved in the crossover is given by the crossover rate. A pool of individuals are randomly selected for crossover and then randomly paired with one another. The switching position is randomly chosen for each pair and offspring produced by combining the left part of one parent with the right part of another and vice versa. For example, if the two parents P_A and P_B, with parameter values a_1 to a_{10} and b_1 to b_{10}, are crossed at location 3, two offspring, O_1 and O_2 will be created as shown below.

$$P_A = (a_1 \ a_2 \ a_3 \ a_4 \ a_5 \ a_6 \ a_7 \ a_8 \ a_9 \ a_{10}) \qquad P_B = (b_1 \ b_2 \ b_3 \ b_4 \ b_5 \ b_6 \ b_7 \ b_8 \ b_9 \ b_{10})$$
$$O_1 = (a_1 \ a_2 \ a_3 \ b_4 \ b_5 \ b_6 \ b_7 \ b_8 \ b_9 \ b_{10}) \qquad O_2 = (b_1 \ b_2 \ b_3 \ a_4 \ a_5 \ a_6 \ a_7 \ a_8 \ a_9 \ a_{10})$$

3.2.5.2 Multipoint Crossover

The simple crossover depends on the order of parameters within an individual and is good at retaining important relationships between adjacent parameters. Nevertheless, as the recombination is dependant on the order of the parameters, many possibly useful recombinations cannot be obtained. For example, when considering a structural identification problem with a total of n mass variables (m_i) and stiffness variables (k_i), it may be natural to keep adjacent stiffness values together and as such arrange the individuals as $[k_1 \ k_2 \ldots k_n \ m_1 \ m_2 \ldots m_n]$. Just as importantly, however, the corresponding pairs of mass and stiffness, for example it may be good to keep $k_1 \ m_1$ together. If simple crossover is used the order of the parameters is important and finding an appropriate arrangement may be difficult.

The multipoint crossover aims to overcome this drawback by recombining parameters with no dependence on the order. While this increases the chance of some potentially useful combinations being destroyed, as crossover points can occur at any number of locations, it does allow for *any* combination of parameters to survive. The number of individuals involved in crossover for a given generation is again controlled by the crossover rate and pairs of individuals are randomly selected for crossover. The multipoint crossover uses many switching positions, allowing recombination of parameters from any location in the individuals. The crossover is performed by considering each parameter in turn. A random number in the range [0 1] is generated and crossover of the parameter is performed when a value greater than 0.5 is returned. An example of this recombination is shown below where the random numbers generated are (0.12, **0.98**, **0.76**, 0.43, 0.23. 0.01, **0.63**, 0.46, 0.36, **0.81**) resulting in crossover at the 2nd, 3rd, 7th and 10th parameters.

$$P_A = (a_1\ a_2\ a_3\ a_4\ a_5\ a_6\ a_7\ a_8\ a_9\ a_{10}) \qquad P_B = (b_1\ b_2\ b_3\ b_4\ b_5\ b_6\ b_7\ b_8\ b_9\ b_{10})$$

$$O_1 = (a_1\ b_2\ b_3\ a_4\ a_5\ a_6\ b_7\ a_8\ a_9\ b_{10}) \qquad O_2 = (b_1\ a_2\ a_3\ b_4\ b_5\ b_6\ a_7\ b_8\ b_9\ a_{10})$$

While the crossover appears to be highly disruptive at first, it must be kept in mind that the order of parameters here does not have the same influence as bit ordering in binary coded GA, where adjacent bits contribute to the coded parameter value by similar amounts. The effect on a real coded GA must be considered in terms of parameter combinations. It is true that in the above example the combinations such as $(a_1\ a_2)$, and $(b_6\ b_7)$, which would most likely survive a simple crossover, are destroyed. We also note however, that other parameter combinations such as $(a_2\ a_3\ a_7\ a_{10})$ are preserved, a result that would be highly unlikely in a simple crossover.

In the iGAMAS, both forms of crossover are used. The rationale is that some reasonable ordering of parameters is usually possible, but at the same time it is not desirable to restrict the algorithm to selecting only the parameter combinations allowed by the given ordering. The two forms of crossover are applied one after the other. Where both simple and multipoint crossovers are to be used, the total crossover rate should be considered. If a crossover rate of P_{cs} is used for simple crossover, and P_{cm} for multipoint crossover, the effective total crossover rate, P_{ct} which is the chance of an individual being involved in *at least one* crossover is,

$$P_{ct} = 1 - (1 - P_{cs})(1 - P_{cm}) \tag{3.6}$$

3.2.6 *Fitness Evaluation and Selection*

For the dynamic problems encountered in this book, fitness may be evaluated from the total sum of square error (SSE) between the simulated and measured response of the system. At each time step the error between the measured and simulated data is computed and squared. The sum of all these errors over all measured degrees of freedom is returned as the SSE. In the programs developed the fitness is evaluated using a bounded fitness function as

$$Fitness = \frac{1}{0.001 + SSE} \tag{3.7}$$

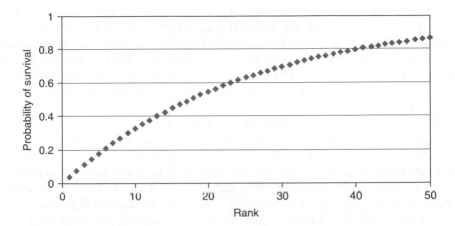

Figure 3.8 Survival probabilities for a population of 50 individuals

This function bounds the maximum fitness at 1000 as errors approach 0 giving us an indication of the extent to which the results have converged. Generally selection would then be carried out by allocating a selection probability to each individual based on its fitness. It is noted however that, as the identification proceeds, many individuals may have very similar fitness values and the selection procedure becomes almost random. To avoid this problem a ranking procedure is used to determine the selection probabilities within each species. The individuals are ranked with the worst individual assigned a rank of 1, the next worst a rank of 2 and so on. The best is thus assigned a rank equal to the population size. Reproduction is then carried out by the commonly used roulette wheel method whereby an individual's chance of selection is proportional to its rank as shown in equation 3.8. This procedure ensures that the fittest individual will always have twice the chance of selection of an average individual within each species. The probability of survival (selection at least once) of an individual with rank R, in a population of size n, can be computed as shown in equation 3.9.

$$P_{selection}(R) = \frac{R}{\sum R} = \frac{2R}{n^2 + n} \tag{3.8}$$

$$P_{survival}(R) = 1 - P_{\text{no selection}}(R) = 1 - \left(\frac{n^2 + n - 2R}{n^2 + n}\right)^n \tag{3.9}$$

The survival probability for the best individual and median individual as n becomes large, can be easily computed as 0.865 and 0.632 by substituting R with n or $(n+1)/2$ respectively, and noting that the asymptotic limits of $(1 + 1/n)^n$ and $(1 - 1/n)^n$ are e and $1/e$ respectively. The convergence to these values is very rapid and even for a population size of only 9 the survival probabilities are within 3% of the limiting values. A plot of the survival probabilities for a population of 50 is shown in figure 3.8.

While the survival probabilities for the best solutions may at first appear low, three factors are to be considered pertaining to selecting the best solutions. The first factor is that in any given population there are likely to be multiple selections of the best results from previous generations. This means that while the best solution has a 14% chance of not being selected, the 2nd best may represent the same, or a very similar solution. The second factor of consideration is that we do not want to saturate the species with the same solution and so it is crucial that the selective pressure on the best individuals is not set too high. Finally, the survival of the best results over all species and generations is guaranteed by the artificial selection procedure, ensuring the elite *will* survive.

As the name of the method implies, artificial selection is crucial to the functioning of the iGAMAS. Artificial selection involves ensuring that the fittest individuals generated over all of the species are stored in species 1 for future refinement by species 4. This simple procedure involves comparing the fitness value of the weakest individual in species 1 with the fitness of the individuals in species 2, 3 and 4. If any of the solutions represent an improvement over those in species 1 they replace them so that species 1 always contains the best solutions that have been obtained. The original (raw, not ranked) fitness values must of course be used to ensure valid comparison of individuals across different species.

One potential problem with artificial selection is that the same individual could be selected many times, thereby saturating species 1. To eliminate this possibility and ensure diversity is maintained, a new idea of tagging is proposed. The tagging guarantees diversity by blocking multiple selections of the solutions with the procedure as follows:

- All individuals are initially assigned a 0 tag.
- If an individual is selected for species 1 its tag is changed to 1.
- The tag follows the individual wherever it goes, through migration, selection and reintroduction.
- If an individual is altered in any way through mutation, crossover or regeneration it no longer represents the same solution and its tag is thus changed back to 0 making it available again for selection to species 1.

3.2.7 Reduced Data Length Procedure

For identification of dynamic systems the simulated response of the system must be calculated for comparison with the measured values. This is of course the most computationally demanding part of the whole process and is responsible for most of the time used. To improve computational efficiency a reduced data length procedure is proposed. The idea is to use a small portion of the total available data to roughly identify the parameters before increasing to the full data set later in the process. Using a reduced initial data length may also help in identification success, as the shorter length gives rise to a smaller number of local optima, thereby increasing the possibility of discovering the desired solution. In the iGAMAS, the procedure includes specifying a cut-off point where the evaluation switches from reduced to full data. The cut-off point and the length of the reduced data to use again depend on the problem but an indication is given in the examples presented in the following chapters. In general, if noise is present, a longer reduced data sequence may be required to help average out the effect of the noise. This procedure is designed for time domain comparisons but could also be used for comparison in the frequency domain. For example, if the objective is to match the frequency response spectrum, a limited number of points in the response spectrum could be used first before the detailed spectrum is calculated later in the analysis.

3.3 Chapter Summary

A novel GA based identification strategy has been presented in this chapter. The strategy is a search space reduction method (SSRM) which utilises the search capability of an improved GA based on migration and artificial selection (iGAMAS). The strategy is described as a two-tier approach in which the SSRM uses the results of the iGAMAS to reduce the search space and to return new search limits to the iGAMAS for further identification.

The SSRM is intuitive in its design, and has been illustrated with a simple example. By reducing the search space of parameters that converge quickly, we are not only able to increase the accuracy of these parameters, but are also better in identifying other parameters, the variation of which now has a relatively larger influence on the objective function. The iGAMAS provides a robust search, simultaneously allowing for broad search while preserving and improving the most promising individuals. The population is split into multiple species, with real encoding of variables and appropriate mutation operators, controlling the search direction. Rank based selection is used to maintain a constant selective pressure, while a tagging procedure guarantees diversity in the pool of best solutions. A reduced data length procedure further improves performance by allowing rapid completion of early generations.

The identification strategy has been presented in general terms in this chapter as it is designed to be easily applied to a wide range of problems, be it financial, mathematical, biological, structural, hydrodynamic, etc. As long as the system can be represented by a reasonably accurate numerical model, capable of reproducing the systems response to a given input, the SSRM can be used. This simple application is a feature of GA based algorithms and is displayed in the following chapters, where the SSRM is applied to specific problems.

Chapter 4

Structural Identification by GA

In this chapter the knowledge needed to develop successful structural identification strategies using genetic algorithms is progressively developed. The concepts behind structural identification are first described using a simple single degree of freedom oscillator and a classical GA. Application then moves to more complex structures of multiple degrees of freedom. The improved GA strategy described in the previous chapter, referred to as simply SSRM (which includes the use of iGAMAS) is used to identify unknown stiffness and damping parameters. In order to properly assess the capabilities of the SSRM, comparison of results with a random search algorithm and a simple GA (SGA) is provided for the known mass systems. It is shown that, even for these relatively simple cases, the SSRM provides far superior identification results. The much more difficult case of identifying systems where the mass is also unknown is then considered before a final extension is made in the following chapter to output-only problems where the stiffness and damping properties are identified without measurement of input excitation. The output-only procedure also estimates the input forces as the identification proceeds.

Throughout this book, it must be kept in mind that the end goal is procedures that not only perform well in theory, but that may be applied to real systems. In this chapter, the numerical strategy is tested in increasing complexity: from one to many degrees of freedom, and from known mass system to unknown mass system. In addition, the practical issues of measurement noise effect and high computational cost are also addressed. In developing the SSRM for use in practical structural problems the following points are considered.

- The method should not require an unreasonably good *initial guess* of the parameters in order to converge.
- Real I/O measurements contain noise and the method should be tested in the presence of *I/O noise*.
- The method should operate on *incomplete measurements* as it is not practical to have measurements at all degrees of freedom in a structure.
- Dynamic measurements are usually obtained using accelerometers and numerical error is inevitable in integration of acceleration to compute velocity and displacement. It is therefore preferable to *utilise accelerations directly* for the identification procedure.

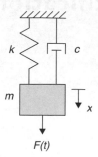

Figure 4.1 SDOF System

4.1 Applying GA to Structural Identification

In order to illustrate the problem of structural identification we will first consider the single degree of freedom (SDOF) system shown in figure 4.1. The equation of motion for the system is given in equation 4.1. For the identification problem here we assume the mass m and the damping c of the oscillator are known and that we want to determine the stiffness k using the measured input force F and acceleration response $\ddot{x}$.

$$m\ddot{x} + c\dot{x} + kx = F \tag{4.1}$$

In order to illustrate this problem, we will use the same simple GA described in Chapter 2 and illustrated in figure 2.2. The only modification required in order to use the simple GA from chapter 2 is an alteration of the fitness calculation to include a dynamic simulation. In order to simulate the dynamic response Newmark's constant average acceleration method is adopted. The response is computed using the given trial parameters using the following steps.

1. The system is assumed initially to be at rest. $x = 0$, $\dot{x} = 0$ and $\ddot{x} = 0$
2. At each time step the incremental displacement Δx is computed according to equation (4.2).

$$\Delta x = \left[\frac{4}{h^2}m + \frac{2}{h}c + k \right]^{-1} \left[F_{k+1} + m\ddot{x}_k + \left[c + \frac{4}{h}m \right]\dot{x}_k - kx_k \right] \tag{4.2}$$

3. The displacement and velocity are then computed from the incremental displacement

$$x_{k+1} = x_k + \Delta x \tag{4.3}$$

$$\dot{x}_{k+1} = \frac{2\Delta x}{h} - \dot{x}_k \tag{4.4}$$

4.   Finally, the acceleration is computed by enforcing dynamic equilibrium at $k+1$, and the procedure is repeated from step 2 until the entire time history is complete.

$$\ddot{x}_{k+1} = m^{-1}(F - c\dot{x}_{k+1} - kx_{k+1}) \tag{4.5}$$

Once the response is simulated it can be compared to the measured response and the fitness computed from the sum of square error as described in equation 3.7. The idea is that, parameters that produce a response close to the measured response should be close to the actual parameters of the system. As the errors become smaller, the parameters will converge to the actual system parameters. An example code of how this simulation and fitness computation can be included is shown below, and a full program for SDOF identification is included in the appendix. The procedure given in the box below must be included in a loop to compute the fitness for each of the individuals within the population.

Simulation and fitness computation for SDOF system

```
x=0
v=0
a=0
```

```
DO t=1,L
   delx=(F(t)+m*a+(c+4*m/h)*v-k*x)/(4*m/(h*h)+2*c/h+k)
   x=x+delx
   v=2*delx/h-v
   a=(F(t)-c*v-k*x)/m
   a_s(t)=a
END DO
SSE=SUM((a_s-a_m)**2)
Fit=1/(0.001+SSE)
```

4.1.1   Sample Problem

In order to demonstrate the strategy described above an example system of $m=1\,\text{kg}$ $k=1\,\text{kN/m}$ and $c=1\,\text{Ns/m}$ (approximately 1.6% critical damping) is considered. The system is excited with a force as given in equation 4.6 and the response simulated for one second at a time step of 0.005 s. The input force and resulting acceleration response is shown in figure 4.2.

$$F = \sin(6\pi t) + \sin(15\pi t) \tag{4.6}$$

Assuming search limits of 500 to 2000 N/m for the unknown stiffness and requiring a resolution of at least 1 N/m results in a string of 11bits. The program is run 25 times using random initial populations of size 10, crossover rate of 0.8, mutation rate of 0.05 and 50 generations. In 21 of the 25 runs, the identified stiffness was 999.7557 N/m, which is the nearest value to the exact value of 1000 that can be represented by the binary coding used. In all of the 25 runs the worst result obtained was 997.5574,

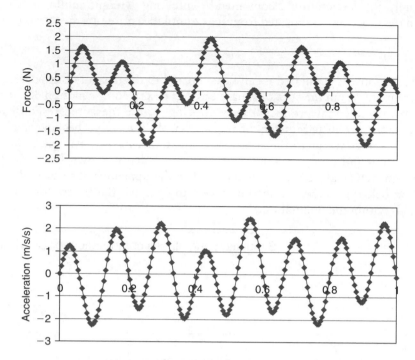

Figure 4.2 Force and response used for sample problem

less than 0.25% from the exact solution. While this example illustrates the ease at which a GA can be applied to structural identification, real structures are more than a single DOF. Due to increased complexity, the simple GA struggles on the multiple-DOF problems and it is therefore necessary to use the SSRM to identify these systems.

4.2   Identification of MDOF Systems Using SSRM

The extension from one to many DOFs is conceptually simple; however, in reality the increase in the size of the search space results in a huge increase in the problem complexity. In the previous section, a single stiffness value was identified using an 11 bit binary sequence, resulting in a search space of 2048 possible solutions. If the same resolution is then used to identify a problem with 10 unknown stiffness values the number of points in the search space jumps to a massive 1.29×10^{33}. In this section the extension to MDOF structures is discussed and the strategy is applied to identify structures where the mass may be known or unknown prior to identification. An important investigation into the effects of noise and data length is also presented.

The structural systems considered in this chapter are two-dimensional shear buildings. The structures consist of rigid beams and flexible columns, reducing the motion to a single translational degree-of-freedom at each floor level as shown in figure 4.3.

The system dynamics are modelled as given in equation 4.7, where the mass, stiffness and damping matrices are readily formed from the structural properties as shown in equations 4.8 to 4.10 respectively. Equation 4.7 is identical to equation 4.1 used for

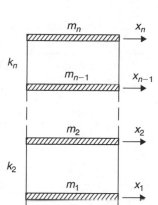

Figure 4.3 n-DOF Structure

the single DOF system, except now the mass stiffness and damping are represented by matrices and the response is a vector representing the motion at the respective floor levels. The mass of the structure is lumped at each floor level and damping is provided as Rayleigh damping where the damping ratio (ζ) is set as 5% in the first two modes of vibration by selecting appropriate values for α and β. The matrices are all banded and constant over time allowing for an efficient numerical procedure to be developed. Newmark's constant average acceleration integration scheme using an efficient LU factorisation of the matrices is used to carry out the simulation of the structural response to a given excitation. More details about Newmark's method and the LU factorisation scheme used are given in the appendix.

$$\mathbf{M\ddot{x} + C\dot{x} + Kx = F} \tag{4.7}$$

$$\mathbf{M} = \begin{bmatrix} m_1 & & & 0 \\ & m_2 & & \\ & & \ddots & \\ 0 & & & m_n \end{bmatrix} \tag{4.8}$$

$$\mathbf{K} = \begin{bmatrix} k_1 + k_2 & -k_2 & & & 0 \\ -k_2 & k_2 + k_3 & -k_3 & & \\ & -k_3 & \ddots & \ddots & \\ & & \ddots & k_{n-1} + k_n & -k_n \\ 0 & & & -k_n & k_n \end{bmatrix} \tag{4.9}$$

$$\mathbf{C} = \alpha \mathbf{M} + \beta \mathbf{K}; \quad \varsigma_r = \frac{\alpha}{2\omega_r} + \frac{\beta \omega_r}{2} \tag{4.10}$$

Table 4.1 Structural properties

	5-DOF	10-DOF	20-DOF
Stiffness ($\times 10^6$ N/m)	(1–2) 400	(1–4) 500	(1–10) 500
(Levels)	(3–5) 250	(5–8) 400	(11–15) 400
		(9–10) 300	(16–20) 350
Mass ($\times 10^3$ kg)	(1–3) 500	(1–5) 600	(1–10) 400
(Levels)	(4–5) 400	(6–10) 420	(11–20) 300
Damping (5% in first 2 modes)			
Alpha	0.573346	0.344092	0.215140
Beta	0.003468	0.005812	0.009224
Natural Period (s)			
First mode	0.796	1.321	2.123
Second Mode	0.300	0.505	0.797

Table 4.2 Location of forces and measurements

	Levels
5-DOF System	
Forces applied	5
Acceleration measurements – mass known	2, 5
Acceleration measurements – mass unknown	1, 3, 5
10-DOF System	
Forces applied	5, 10
Acceleration measurements – mass known	2, 4, 7, 10
Acceleration measurements – mass unknown	1, 2, 4, 6, 8, 10
20-DOF System	
Forces applied	5, 10, 15, 20
Acceleration measurements – mass known	2, 4, 7, 10, 12, 14, 17, 20
Acceleration measurements – mass unknown	1, 2, 3, 4, 6, 8, 10, 12, 14, 16, 18, 20

Shear buildings of 5, 10 and 20-DOF are considered, with structural properties given in table 4.1. For each structure, the mass and stiffness properties are first decided. The eigenvalues of the system are then computed and the natural frequencies obtained are used to determine the values of α and β so as to provide the required damping ratio of 5% in the first two modes.

Excitation is provided as random white Gaussian noise (WGN) input forces, scaled to have a root-mean-square (RMS) value of 1000 N. The response of the structure is then simulated and the accelerations recorded for feeding into the SSRM. For the tests conducted in sections 4.2.1 and 4.2.2, simulation is carried out for 200 data points at a time step of 0.01 s. The forces are applied at every 5th floor and acceleration measurements are obtained from the simulations at selected floors as given in table 4.2. As the identification of systems where the mass is unknown is more difficult, more acceleration measurements are used than for the known mass case.

The SSRM is first applied to the MDOF problem by arranging the structural parameters into a vector as shown in figure 4.4. The vector contains all of the stiffness, mass

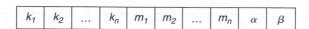

Figure 4.4 Parameter vector used for *n*-DOF system

and damping parameters of the structure. In order to obtain a feel of the appropriate GA parameters for the SSRM and iGAMAS, the investigation first looks at known mass systems without noise in section 4.2.1. For problems where some of the parameters are known, for example in known mass problems, the search limits for those parameters can simply be input as the exact values. The investigation then moves to the more difficult unknown mass problems in section 4.2.2, before we finally introduce input and output noise and the effects of available data length and the benefits of the reduced data length procedure in section 4.2.3.

    For all tests conducted in this chapter, identification is carried out 25 times using fresh input forces and noise and a summary of the results is presented.

4.2.1 *Known Mass Systems*

The case of known mass shear buildings is used to compare the performance of the SSRM, the iGAMAS alone, a simple GA (SGA) and a random search algorithm. The SGA used here differs from that used for the SDOF example in that a real encoding of the parameters is used. Mutation is provided as random regeneration of the selected parameter. When conducting tests to compare identification strategies, two approaches are possible. The first is to compare the time taken in achieving a given accuracy, while the second is to compare the accuracy that can be achieved in a given time. In all of the examples presented in this book, the latter comparison method is used by fixing the total evaluations. The total evaluations refer to the number of times the time history simulation (forward analysis) is carried out and are set here as 10,000, 20,000 and 80,000 for the 5, 10 and 20-DOF systems respectively. The corresponding computational times are approximately 3 s, 12 s, and 100 s when the simulation is conducted on a standard Pentium 4, 3-GHz PC. The number of evaluations may seem large but is actually only a very small portion of the entire search space. Consider for example the 20-DOF system (with 22 unknowns). If we were to partition each variable into 150 sections, representing 1% resolution of the parameter value, there would be approximately 150^{22} ($\sim 10^{48}$) regions in the search space to be evaluated. To evaluate every point on this search space using the same computer used in these tests would require 3×10^{37} years! The 80,000 evaluations used here therefore represent only a tiny fraction of the search space.

    With any GA it is important to determine balanced parameters whereby the combination of population sizes, number of generations, mutation and crossover probabilities, etc are able to work well together to produce consistently good results without requiring excessive computation time. In preparation for this section, a preliminary study was conducted in order to establish good GA parameters to use for each of the algorithms. In all cases the population sizes, number of generations, mutation and crossover rates etc were varied in order to determine parameters that gave consistently good results. This study was essential in obtaining a feel for the GA parameters that would work well for the SSRM and iGAMAS algorithms. The GA parameters determined from

Table 4.3 Known mass systems – GA parameters

	5-DOF	10-DOF	20-DOF
SSRM			
Population size	7×3	9×3	19×3
Runs**	4/9	4/9	4/9
Generations	53	82	156
Crossover rate	0.8	0.8	0.8
Mutation rate	0.2	0.2	0.1
Regeneration	2	2	3
Reintroduction	25	30	42
iGAMAS			
Population size	7×3	9×3	19×3
Generations	476	741	1404
Crossover rate	0.8	0.8	0.8
Mutation rate	0.2	0.05	0.05
Regeneration	5	2	3
Reintroduction	25	30	8
SGA			
Population size	80	113	226
Generations	125	176	354
Crossover rate	0.64	0.96	0.96
Mutation rate	0.05	0.05	0.05

** The first number is the number of runs used in evaluation of limits, and the second number is the total number of runs (see Section 3.1.1)

Table 4.4 Known mass systems – Identification results

	SSRM	iGAMAS	SGA	RANDOM
5 DOF				
Mean error in stiffness	0.43% (0.05)	1.35% (0.12)	4.58% (0.42)	10.5% (1.0)
Mean error in damping	2.27% (0.40)	6.18% (1.08)	9.41% (1.21)	28.55% (3.29)
10 DOF				
Mean error in stiffness	0.43% (0.03)	1.36% (0.10)	4.22% (0.29)	20.1% (1.10)
Mean error in damping	1.56% (0.28)	5.68% (1.03)	12.33% (1.89)	27.65% (2.37)
20 DOF				
Mean error in stiffness	0.52% (0.03)	2.29% (0.15)	8.33% (0.39)	24.3% (0.9)
Mean error in damping	0.75% (0.17)	4.93% (1.00)	15.81% (2.81)	27.26% (4.11)

* Standard error of the mean values are given in ( )

this preliminary study are shown in table 4.3. It is important to note that the GA parameters are selected to be as consistent as possible across all of the three shear buildings studied. The reduced data length procedure is not used at this stage. Using the GA parameters specified, identification is carried out 25 times for each of the systems studied. The identification results are shown in table 4.4 and a comparison of the mean error in the identified stiffness values is shown in figure 4.5. The results clearly show that the SSRM performs better than the iGAMAS alone, and far better than a simple GA or a random search. Although these results refer to the easier, known mass

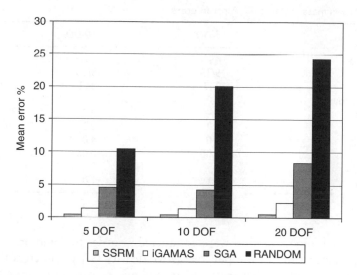

Figure 4.5 Mean error in identified stiffness values

systems and there is no noise in the signals, the accuracy achieved is very encouraging. Furthermore the computational time required to identify the 7 unknown parameters of the 5-DOF system is only 3.2 s. A time of 100 s in identifying 22 unknowns in the 20-DOF system is very reasonable.

4.2.2 Unknown Mass Systems

Unknown mass systems present a far greater challenge than the known mass systems considered in the previous section and have rarely been considered in other structural identification studies. The problem of identifying both mass and stiffness is difficult as, not only are the number of unknowns increased, but different combinations of mass and stiffness can produce the same natural frequencies and mode shapes, leading to similar response characteristics. This fact can be easily illustrated by considering two SDOF systems. The first system has mass of 1 kg and stiffness of 400 N/m, and the second has mass of 5 kg and stiffness of 2 kN/m. It is easily noted that both systems have a frequency of 20 rad/s and would display the same free vibration characteristics. Only by considering forced oscillations can these components be separated and identified. Few studies have attempted to identify mass in structural systems, as the objective is generally to identify damage which is based on changes in stiffness values. In some cases, however, accurate calculation of mass is not possible, particularly when the mass is to be modelled as lumped values. In these cases identification of mass can help in obtaining more realistic estimates of stiffness.

The same three structures are considered in this section, except that now the mass properties are not input exactly but are to be identified in the range of half to twice the actual values. As these problems present a far greater challenge, the total evaluations are increased to 500,000, 1,000,000 and 2,000,000 for the 5, 10 and 20-DOF systems respectively. The computational times are approximately 2 m 40 s, 10 m 30 s, and 42 m for analysis conducted on a standard Pentium 4, 3-GHz PC. As with the previous

Table 4.5 Unknown mass systems – GA parameters

	5-DOF	10-DOF	20-DOF
Population size	45×3	65×3	90×3
Runs	5/15	5/15	5/15
Generations	247	342	494
Runs to ave for output	1	1	1
Crossover rate	0.4	0.4	0.4
Mutation rate	0.2	0.2	0.1
Window width	4.0	4.0	4.0
Migration	0.05	0.05	0.05
Regeneration	3	3	3
Reintroduction	50	120	200

section some preliminary studies were conducted in order to determine some reasonable GA parameter values. The purpose is to identify GA parameters that will give consistently good results rather than trying to obtain so called optimum GA parameters which do not exist in reality due to variation across systems, noise etc. We do not want to change parameters to achieve only a small gain but would rather use parameters that work well over all systems. The GA parameters selected for use in this section are presented in table 4.5.

As with the unknown mass systems of the previous section, the consistency of the GA parameters across the three systems is excellent. The most significant changes, as compared to the known mass case, are the reduction in crossover rate and an increase in the number of runs.

The decrease in crossover rate is logical because of the way the identification strategy is implemented. For the known mass case, the same parameter vector is used but with the limits for mass set to the exact parameter value. This effectively reduces the crossover rate as many crossovers occurring in the mass portion of an individual will have little effect. For the unknown mass case however, all crossovers provide useful recombination of the parameters. In this way we can see that the rate of 0.8 used in the known mass systems and the 0.4 rate used here actually provide a similar number of useful recombinations.

The second important change in the GA parameters is the number of runs. The runs for all systems are increased to 5 runs with total runs of 15. This increase can be explained from two considerations. Firstly, as the unknown mass problems pose a greater challenge, increasing the number of runs leads to better certainty in the mean and a more robust solution. Secondly, as the total evaluations are increased for the unknown mass problems, the runs can be increased without reducing the number of generations too far. For the known mass problems, where the total evaluations are very limited, using 15 total runs causes the number of generations to become very small and the performance of iGAMAS is compromised.

The higher reintroduction rate required by the 10 and 20-DOF systems has confirmed the earlier findings that a high reintroduction is desirable. The results indicate that this high reintroduction rate is crucial to the performance of the strategy and the local search species 4 is working well in refining the search results.

Table 4.6 Unknown mass systems – Identification results

	5-DOF	10-DOF	20-DOF
Mean error – stiffness	0.41 (0.07)	0.64 (0.10)	0.28 (0.03)
Mean error – mass	0.40 (0.07)	0.56 (0.09)	0.32 (0.03)
Mean error – damping	0.97 (0.39)	1.84 (0.52)	1.09 (0.33)

* Standard error of the mean values are given in ( )

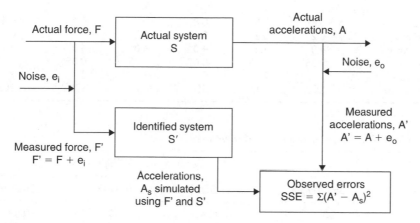

Figure 4.6 Effect of noise on identification

A summary of the achieved accuracy is given in table 4.6. The identification of unknown mass systems is considered very difficult and as such, the results presented here are considered excellent. The next question of course is the quality of identification results when the signals are contaminated with noise. This is investigated in the following section.

4.2.3 *Effects of Noise and Data Length*

In this section the effects of the following two important issues are examined for both known mass and unknown mass systems:

(a) Measurement noise which is inevitable in SHM of real structural systems.
(b) Data length used in the forward analysis which affects the overall computational cost of GA-based structural identification.

The same systems are considered but the input forces and measured accelerations are contaminated with noise. Applying noise to both the inputs and outputs is a much more general and difficult case compared to the common case of only output noise. The effect of this noise is illustrated in figure 4.6, where it is seen that the noise on the force is passed through the simulation and causes errors in the simulated accelerations, which are used to compute the fitness value. All tests are again carried out 25 times. For each test, noise is freshly generated to avoid any bias that might result from a single noise pattern.

Table 4.7 GA parameters for study on the effect of noise and data length

	Known Mass Systems			Unknown Mass Systems		
	5-DOF	10-DOF	20-DOF	5-DOF	10-DOF	20-DOF
Total Evaluations	10,000	20,000	80,000	500,000	1,000,000	2,000,000
Population size	7 × 3	9 × 3	19 × 3	45 × 3	65 × 3	90 × 3
Runs	4/10	4/10	4/10	5/15	5/15	5/15
Generations	48	74	140	247	342	494
Runs to ave for output	1	1	1	1	1	1
Crossover rate	0.8	0.8	0.8	0.4	0.4	0.4
Mutation rate	0.2	0.2	0.1	0.2	0.2	0.1
Window width	4.0	4.0	4.0	4.0	4.0	4.0
Migration	0.05	0.05	0.05	0.05	0.05	0.05
Regeneration	2	2	3	3	3	3
Reintroduction	25	30	50	50	120	200

Table 4.8 Effect of noise and data length – Known mass

		Mean error in identified stiffness %		
Noise level	Data Length	5-DOF	10-DOF	20-DOF
0%	50	0.54	0.99	0.69
	100	0.25	0.30	0.39
	200	0.44	0.26	0.44
	500	0.31	0.45	0.38
	1000	0.50	0.45	0.44
5%	50	9.40	5.32	5.55
	100	2.64	3.27	3.33
	200	1.96	1.90	2.34
	500	1.06	1.28	1.61
	1000	1.07	0.92	1.11
10%	50	17.61	10.45	11.92
	100	5.59	5.82	6.86
	200	3.88	3.61	4.40
	500	2.40	2.34	2.70
	1000	1.36	1.69	2.32

Three noise levels are considered, namely 0%, 5% and 10% of the given signals based on the RMS values. For each case, data lengths of 50, 100, 200, 500 and 1000 points are considered and the resulting accuracy and computational times are compared. In all cases the GA parameters are as shown in table 4.7, based on the findings of the previous sections.

A summary of the results obtained is presented in tables 4.8 and 4.9 and the variation of the error in stiffness illustrated in figure 4.7. The results for errors in identified mass are similar to those for stiffness and are not discussed here. When studying the results it must be kept in mind that the GA parameters are exactly the same in all cases, and as such, the time taken is proportional to the data length.

Table 4.9 Effect of noise and data length – Unknown mass

Noise level	Data Length	Mean error in identified stiffness		
		5-DOF	10-DOF	20-DOF
0%	50	1.74	0.72	1.88
	100	0.62	1.13	0.62
	200	0.41	0.64	0.28
	500	0.90	0.81	0.29
	1000	0.78	0.56	0.25
5%	50	8.83	6.09	5.84
	100	4.68	3.50	3.33
	200	2.34	2.29	2.02
	500	1.73	1.59	1.21
	1000	1.25	1.38	0.98
10%	50	16.36	13.11	11.98
	100	7.25	7.09	6.39
	200	5.69	4.43	4.44
	500	2.55	2.88	2.65
	1000	2.11	2.41	2.08

The noise causes the error in identified parameters to increase approximately in proportion to the noise level applied. Nevertheless the results are still excellent in general, as even under a large 10% noise, stiffness and mass properties are identified with good accuracy.

The effect of the data length is twofold. Firstly, a longer data length provides more information and helps to 'average out' the effect of noise such that a better identification is possible. It is conceivable that there is also a second effect, whereby using more data points may make it harder to find the desired solution due to many local optima created, leading to a slower convergence of results. This effect is observed in the case of zero noise where the best results are achieved at a smaller data length.

In general, for realistic noisy data, the first effect will dominate and using more data should help to improve the quality of the identification. The combination of these two effects does, however, support the idea behind the reduced data length procedure. That is, we can use a shorter length at first to achieve a faster convergence and then increase to a longer data length later in order to 'fine tune' the result and reduce the effect of noise. This reduced data length procedure is developed and illustrated in the following section.

4.2.3.1 Reduced Data Length Procedure

The primary purpose of the reduced data length is to complete more evaluations in a shorter time and hence improve the quality of the identification. For the tests presented in this section, the total time is fixed as that taken for full runs of 200 data points, with the result from the previous section taken as a basis for comparison. The computational times for the known mass case are approximately 3.2 s, 12.4 s and 100 s (on a Pentium 4, 3 GHz computer) for the 5, 10 and 20-DOF systems respectively. The corresponding

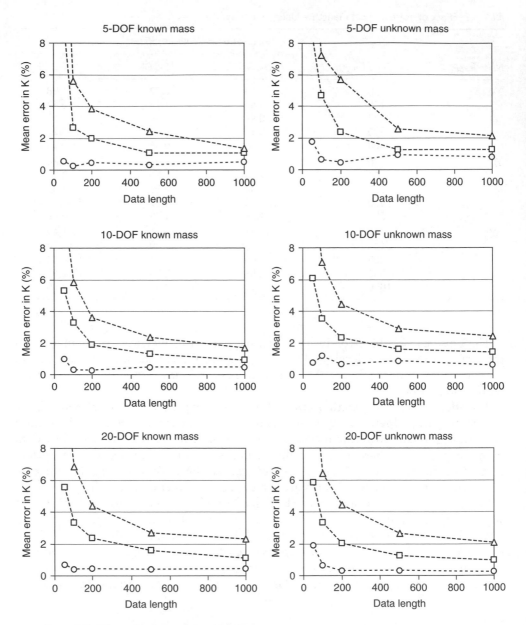

Figure 4.7 **Effect of noise and data length**
Circle = 0% noise, square = 5% noise, triangle = 10% noise

times for the unknown mass case are 2 min 40 s, 10 min 30 s and 42 min. As the computational time taken is approximately proportional to the data length, the total data points evaluated are fixed and computed as the total evaluations times the data length. The tests are conducted by varying the full data length, the reduced data length, and the percentage of generations for which the reduced length should be used. With

Table 4.10 Reduced data length – Best results

System (time taken)	Noise	Full data length	Reduced data length	% Gen*	Mean % error	
					Stiffness	Mass
Known mass systems						
5-DOF (3.2s)	5%	1000	200	75	1.30 (0.11)	– –
	10%	500	200	50	2.20 (0.20)	– –
10-DOF (12.4s)	5%	500	200	50	1.51 (0.09)	– –
	10%	500	50	50	2.56 (0.15)	– –
20-DOF (100s)	5%	500	100	50	1.59 (0.07)	– –
	10%	500	0	0	3.05 (0.12)	– –
Known mass systems – Double number of generations						
5-DOF (6.4s)	5%	500	200	50	0.94 (0.08)	– –
	10%	1000	100	50	1.72 (0.18)	– –
10-DOF (25s)	5%	1000	200	75	1.17 (0.07)	– –
	10%	1000	200	50	1.82 (0.09)	– –
20-DOF (200s)	5%	1000	200	50	1.33 (0.06)	– –
	10%	1000	200	50	2.35 (0.10)	– –
Unknown mass systems						
5-DOF (2 min 40 s)	5%	500	200	75	1.21 (0.08)	1.18 (0.10)
	10%	500	200	50	2.10 (0.17)	2.31 (0.17)
10-DOF (10 min 30 s)	5%	500	100	50	1.43 (0.07)	1.36 (0.07)
	10%	1000	200	75	2.82 (0.14)	2.86 (0.15)
20-DOF (42 min)	5%#	1000	100	50	1.24 (0.04)	1.61 (0.07)
		500	200	75	1.31 (0.05)	1.41 (0.06)
	10%	1000	100	50	2.33 (0.09)	2.63 (0.09)

* % gen is the percentage of total generations that are run using the reduced data length before analysis switches to full data.
# Two results given here as best mass and stiffness did not occur for same case.

these parameters the number of generations is chosen such that the total number of points evaluated (and hence time taken) will be approximately equal. As the time taken for the known mass problems is reasonably short, trials are also conducted with the number of generations doubled. All other GA parameters remain unchanged from those used in the previous section.

The values used for the trials are full data lengths of 200, 500 and 1000, reduced data of 50, 100 and 200, and percentage of generations to run reduced data of 50, 75 and 90%. A summary of the best results is presented in table 4.10. Herein 'best' refers to the best average performance over 25 tests and not the single best result obtained.

While the best results are impressive, we are ultimately interested in finding some standard parameters and data lengths that deliver consistently good results over all systems. The results for a data length of 500 with a reduced length of 200 used for 50% of the generations (denoted here as 500/200/50) are reasonably good over all systems. Table 4.11 gives a summary of these results and also gives the deviation that the results fall from the best results given in table 4.10. The deviation is given in terms

Table 4.11 Reduced data length – Results for 500/200/50

		Mean % error		Deviation from best*	
System	Noise	Stiffness	Mass	Stiffness	Mass
Known mass systems – normal gen					
5-DOF	5%	1.36 (0.12)	–	0.06 (0.37)	–
	10%	2.20 (0.20)	–	0	–
10-DOF	5%	1.51 (0.09)	–	0	–
	10%	2.78 (0.15)	–	0.22 (1.04)	–
20-DOF	5%	1.81 (0.08)	–	0.22 (2.07)	–
	10%	3.47 (0.15)	–	0.42 (2.19))	–
Known mass systems – 2x gen					
5-DOF	5%	0.94 (0.08)	–	0	–
	10%	2.32 (0.16)	–	0.60 (2.49)	–
10-DOF	5%	1.37 (0.07)	–	0.20 (2.02)	–
	10%	2.47 (0.14)	–	0.65 (3.91)	–
20-DOF	5%	1.76 (0.08)	–	0.43 (4.10)	–
	10%	2.97 (0.12)	–	0.62 (3.97)	–
Unknown mass systems					
5-DOF	5%	1.63 (0.13)	1.61 (0.14)	0.42 (2.69)	0.43 (2.50)
	10%	2.10 (0.17)	2.31 (0.17)	0	0
10-DOF	5%	1.60 (0.09)	1.50 (0.09)	0.17 (1.49)	0.14 (1.23)
	10%	2.98 (0.15)	3.00 (0.15)	0.16 (0.78)	0.14 (0.66)
20-DOF	5%	1.38 (0.05)	1.51 (0.06)	0.14 (2.19)	0.10 (1.18)
	10%	2.78 (0.10)	3.00 (0.11)	0.45 (3.34)	0.37 (2.60)

*The value in ( ) is the deviation given in terms of a number of standard errors as shown in (Eq 4.11).

of the absolute difference in mean error as well as in terms of a number of standard errors as given in equation 4.11. $\bar{x}$ and $\bar{x}_B$ are the mean identification errors obtained using 500/200/50 and best results respectively, while $s_{\bar{x}}$ and $s_{\bar{x}_B}$ are the standard errors.

$$\text{deviation} = \frac{|\bar{x} - \bar{x}_B|}{\sqrt{s_{\bar{x}}^2 + s_{\bar{x}_B}^2}} \tag{4.11}$$

The deviation reported in this way gives a good feel as to how significantly different the results are. While in some cases there is a significant difference between the 500/200/50 result and the best, the accuracy obtained is still very good and for practical purposes the convenience of having a single length is desirable. It is also interesting to compare the results to those obtained using full runs of 500 data points. The mean error in stiffness is on average 0.26% better using the 500/200/50 reduced data length. In six cases the results are significantly better (95% confidence interval), while in only two cases the result is significantly worse. The proposed reduced data length procedure is therefore considered beneficial in most cases.

4.3 Chapter Summary

The effectiveness of the SSRM strategy in identifying the parameters of structural systems has been demonstrated in this chapter. The results presented here are derived from more than 40,000 simulation tests carried out over more than 100 computer-days. This huge amount of data has facilitated a comprehensive comparison of results and developed a good understanding of the various GA parameters. An understanding of the effect of the GA parameters has been achieved through many tests conducted using various combinations of the parameters. It is interesting to note that for a classical GA (e.g. the simple GA used for comparison in this study), large population sizes are generally preferred, whereas small populations tend to work better for the SSRM. The mutation rates are also very different, as large mutation rates are preferred for the SSRM compared to small rates for the simple GA. The total number of runs required for the SSRM is independent of the system size and is found to be about 10 and 15 for known mass and unknown mass systems respectively.

The results have clearly shown that the SSRM performs better than the iGAMAS alone, and far better than a simple GA and obviously a random search. The strategy has been demonstrated on both known and unknown mass systems as well as in the presence of input and output noise. The unknown mass system poses a greater challenge, for which many other identification methods are unable to handle. In the presence of noise, it is generally observed that a longer data length allows for a more accurate identification, but at the expense of additional computation time. The reduced data length procedure allows the combination of fast analysis using a short data length with the accuracy that can be obtained from longer data. It has been shown that, given a fixed time, the reduced data procedure can significantly improve the accuracy of results. The results obtained show that using full data/reduced data/% generations of 500/200/50 appears to work well over all systems, varying at most by 0.6% from the best result obtained using any data length combinations. The results obtained are very accurate with mean errors of less than 2% and 3% achieved for unknown mass systems with noise of 5% and 10% respectively. The proposed strategy is also computationally efficient. A 20-DOF unknown mass system (42 unknowns) is identified in 42 minutes on a standard Pentium 4, 3-GHz PC.

Chapter 5

Output-Only Structural Identification

The structural identification results presented in the previous chapter are very good. In reality, measuring input forces in situations outside the laboratory may not always be feasible. With this in mind, the SSRM is adapted to identify structural stiffness for problems where the input force is not measured. The majority of output-only methods use frequency domain methods. The strengths and weaknesses of these methods were discussed in the first chapter, where the ability to identify the structure without input force information was mentioned as a key benefit. These methods will not be discussed further in this chapter as the global parameters identified by these frequency domain methods are generally not sufficiently sensitive to local damage. By comparison, there are benefits in developing time domain strategies which are able to better identify structural parameters at the local level. While common for frequency based methods, Identification using only output information has rarely been attempted using time domain identification schemes. The most significant works, by Ling and Haldar (2004) used classical techniques to carry out the identification using an iterative procedure. This procedure worked reasonably well but will of course suffer from the same limitations as other classical methods as discussed in the first chapter. Furthermore, the iterative procedure, while reasonably efficient for a least squares identification, would require significant computational time for a GA due to the larger time required for each iteration. In this chapter, a strategy involving *simultaneous* evolution of structural parameters and input force is introduced. This procedure uses the SSRM, but since the input force is unknown, it must also be identified and updated as the search for structural parameters proceeds. In theory, it is mathematically not possible to identify all structural parameters and forces at the same time, as the solutions can "float" by a scalar constant. A similar problem was discussed in section 4.2.2, i.e. the free vibration response of a system would be identical if the mass and stiffness were both scaled by the same factor. The only way to fix the values is to consider forced oscillations with a measured force. If the force is unknown, this becomes impossible as the forces can also be scaled by an arbitrary factor in order to match the mass and stiffness values. We must therefore know at least one of the properties of the system in order to 'fix' the parameters. For the procedure developed here the mass of the structure is assumed known. For many engineering structures, such as bridges or offshore platforms, this is a reasonable assumption as the mass may be estimated with reasonable accuracy.

Before discussing the modifications that are required, let us first properly define the problem to be considered. The aim is to identify the stiffness properties of the structure using only selected, noise contaminated acceleration measurements. The structure is excited with an unknown force time-history. We assume that the structure is initially at rest and that the location of the force(s) is known. The mass of the structure is also assumed to be known, and damping is assumed to be of Rayleigh-type damping where the damping parameters are unknown. The force time history at each location will be identified along with the system parameters. While the identified force history can be also of some interest, it is identified primarily to facilitate the identification of stiffness parameters which is the major objective. It is assumed that acceleration measurements are available at the location where the force is applied, and at adjacent degrees of freedom. Other DOFs may or may not be measured depending on the sensor availability. For example if a 5-DOF shear building is to be identified, and the force is known to act at the 5th level, then it is assumed that measurement is available at least at levels 4 and 5.

5.1 Modification of the Identification Strategy

A procedure involving simultaneous identification of the force and structural parameters is developed. The SSRM is employed as in the previous chapters, except that the force used in the simulations is unknown and has to be computed in order for the simulation of structural response to proceed. The parameter vector used is the same as before and does not include any parameters related to the unknown force. Note that the force is therefore not viewed as a variable to be identified by the GA, but, instead is treated as an unknown component that is required to in the dynamic equilibrium at each time step. In the SSRM the calculation of force is combined with the Newmark's simulation algorithm, resulting in an efficient subroutine that is able to estimate the input force(s) while simultaneously carrying out the simulation of the structural response for comparison with measured accelerations to compute the fitness of the given solution.

The procedure used to compute the force and simulate the response may be thought of as a predictor-corrector algorithm. An initial estimate, or "prediction", of the displacements and velocities at the measured DOFs at time step $k+1$ is first obtained from the measured accelerations at time step $k+1$ and "corrected" response at time step k, using the following "predictor" equations, where h is the time step.

$$\dot{\mathbf{x}}_{k+1} = \dot{\mathbf{x}}_k + \frac{h}{2}(\ddot{\mathbf{x}}_k + \ddot{\mathbf{x}}_{k+1}) \tag{5.1}$$

$$\mathbf{x}_{k+1} = \mathbf{x}_k + \frac{h}{2}(\dot{\mathbf{x}}_k + \dot{\mathbf{x}}_{k+1}) \tag{5.2}$$

Due to the banded nature of the matrices only the response at the loaded DOF and adjacent (coupled) DOFs are needed for computing the force and only the sub-matrix containing the necessary DOFs needs be considered. The unknown force F^u

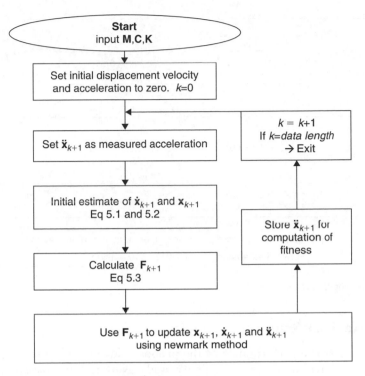

Figure 5.1 Simulation and force calculation procedure

is then computed from the known forces F^{kn}, measured acceleration and predicted displacement and velocity by

$$F^u_{k+1} = M\ddot{x}_{k+1} + C\dot{x}_{k+1} + Kx_{k+1} - F^{kn}_{k+1} \qquad (5.3)$$

Step-by-step integration by Newmark's constant acceleration method is then used to recompute the reponse at time step $k+1$ resulting in revised, or "corrected", estimates of displacement, velocity and acceleration. The corrected accelerations are stored for comparison with the measured accelerations in order to compute the fitness as required in the GA procedure. The corrected response is then passed on to the next time step and the process is repeated for the entire time history as illustrated in Figure 5.1. The key point to note here is that it is the corrected response, rather than the measured or predicted response, that is used as the response vector at time "k" in Eqs. 5.1 and 5.2. The simulation step used to correct the response estimates effectively uses the structural dynamic system to act as a filter, thereby minimizing the effect of high-frequency noise in the measurement. If the corrected (simulated) response is not used, the force is calculated from the response obtained by direct integration of the measured accelerations and as such would be more prone to accumulation of errors. Using the corrected response also ensures that dynamic equilibrium is maintained at the end of each time step.

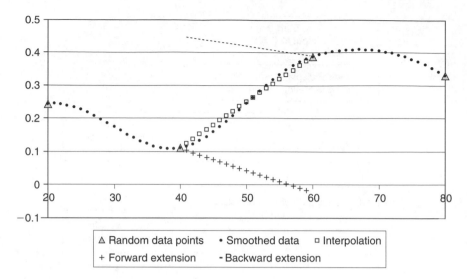

Figure 5.2 Input force generation procedure

5.2 Numerical Study

In order to assess the performance of the proposed identification strategy numerical simulations are carried out. A preliminary study using the proposed strategy shows that a longer data length and small time step may be beneficial. Thus in addition to tests using 500 data points, tests on data lengths of 1000 points are also considered. A time step of 0.001s has been used throughout this section. The reduced data length procedure is used in all cases with a reduced length of 40% of the data points used for 50% of the generations.

As the procedure requires integration of acceleration to obtain velocity and displacement, a very irregular random force should not be used. Rather than a purely random force, a smoothed random wave form is used. The excitation forces must balance having broad frequency content with being smooth enough that they can be accurately integrated for generation of the response in the identification procedure. Here random data are generated every 20 points and the interpolation functions used to fill in the intermediate points, resulting in a sort of 'band limited noise'. In order to create a smoothed random force a random force is first generated at a time step of 0.02 s (sampling frequency of 50 Hz) and the signal is then converted to a time step of 0.001 s (1000 Hz) by interpolating and smoothing the force over the intermediate data points. The wave form in between 2 random points is generated using linear interpolation as well as the backwards and forward extension of the gradients of neighbouring points. The idea is to create a curve that flows smoothly through the random points given. The linear interpolation, backwards and forward extensions are therefore combined using weighted averages to obtain the value of the wave form at intermediate points as illustrated in figure 5.2 for data points 40 to 60. As shown, one of the random data points on either side of the interval of interest is required for the interpolation in order to obtain the forward and backward extensions. The quadratic weighting functions used

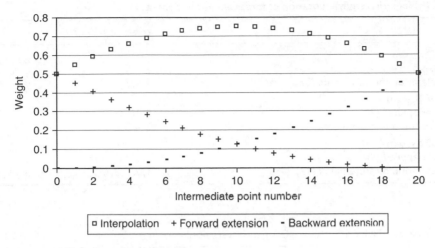

Figure 5.3 Weights for input force generation

are given in equations 5.4–5.6 and illustrated in figure 5.3. Using this method, fresh input forces are prepared for each of the tests conducted. Examples of these forces can be seen in the force comparisons presented later in the section in figures 5.4 and 5.5.

$$w = 0.75 - 0.0025(i - 10)^2 \qquad (5.4)$$

Interpolation weight

Forward extension weight

$$w = 0.00125(20 - i)^2 \qquad (5.5)$$

Backward extension weight

$$w = 0.00125i^2 \qquad (5.6)$$

It is noted here that while the program does not assume any type of function, and a purely random signal could be used, as with any identification scheme, some forms of force input will prove more successful than others. The chosen input forces, while not optimum, do provide a reasonable balance between providing a good excitation and being 'smooth enough' to allow for reasonable numerical integration of the time history responses.

The study considers the same three structures that were used in the previous sections. The location of forces is the same as those used previously; however, due to the method requiring specific acceleration measurements, the response is measured at the degrees of freedom as indicated in table 5.1. Apart from data length discussed above, the GA parameters used are similar to those used previously and are summarised in table 5.2. As before, the search limits of all stiffness and damping parameters are set as half to double the exact values. As identification involves unknown force as well as unknown stiffness and damping parameters, the problem becomes more difficult and additional runs are allowed as compared to the known mass cases where the force was known.

The identification is considered in the presence of 0, 2, 5 and 10% noise. For each case the analysis is repeated 25 times and average results reported. In all cases the stiffness and damping parameters and the force(s) are identified. Errors in identified

Table 5.1 Numerical study – Location of forces and measurements

	Levels
5-DOF System	
Forces applied (1)	5
Acceleration measurements (3)	2, 4, 5
10-DOF System	
Forces applied (2)	5, 10
Acceleration measurements (6)	2, 4, 5, 6, 9, 10
20-DOF System	
Forces applied (4)	5, 10, 15, 20
Acceleration measurements (13)	1, 2, 4, 5, 6, 9, 10, 11, 14, 15, 16, 19, 20

Table 5.2 Numerical study – GA parameters used

	5-DOF	10-DOF	20-DOF
Population size	7×3	10×3	20×3
Runs	5/15	5/15	5/15
Generations	60	100	200
Runs to ave for output	1	1	1
Crossover rate	0.8	0.8	0.8
Mutation rate	0.2	0.2	0.1
Window width	4.0	4.0	4.0
Migration	0.05	0.05	0.05
Regeneration	2	2	3
Reintroduction	25	30	50
Data Length	500 or 1000 with reduced data of 40% used for 50% of generations		

damping are reasonable. As damping in the structures is dominated by the stiffness proportional term, the value for β is well estimated. Even under 10% noise, β is identified with mean error of only 3.02%, 1.12% and 0.53% for the 5, 10 and 20 DOF systems respectively. As α does not contribute as significantly to the damping in these systems it was estimated less reliably. Since our main interest is in the identified stiffness values, the results presented in table 5.3 compare the mean and maximum errors in identified stiffness.

The results presented in table 5.3 are very good. While the time taken for the identification is slightly longer than was used for the known mass problems (with measured force), the identification accuracy is outstanding. The fact that the results, even for 500 data points, are much better than those achieved previously (table 4.11) suggests that noise on force measurement compromises the accuracy when the identification procedure assumes known force (which is commonly done). By identifying force, rather than assuming force measurement to be accurate, we are able to avoid force measurement noise that would otherwise affect the accuracy of the forward analysis needed in the identification procedure. The results also suggest that the smoother force used here is a

Table 5.3 Numerical study – Error in identified stiffness parameters

System and Noise	500 Data Points		1000 Data Points	
	Mean Error	Max Error	Mean Error	Max Error
5-DOF				
Time (min:s)	0:14		0:28	
0% noise	0.13	0.31	0.05	0.12
2% noise	0.44	1.05	0.21	0.50
5% noise	0.93	2.19	0.50	1.20
10% noise	2.24	6.06	0.98	2.31
10-DOF				
Time (min:s)	1:04		2:08	
0% noise	0.05	0.15	0.02	0.06
2% noise	0.33	1.06	0.20	0.59
5% noise	0.84	2.50	0.47	1.23
10% noise	1.70	5.04	0.90	2.70
20-DOF				
Time (min:s)	8:39		17:18	
0% noise	0.01	0.02	0.01	0.04
2% noise	0.32	1.02	0.21	0.66
5% noise	0.78	2.50	0.52	1.55
10% noise	1.43	4.59	0.93	2.91

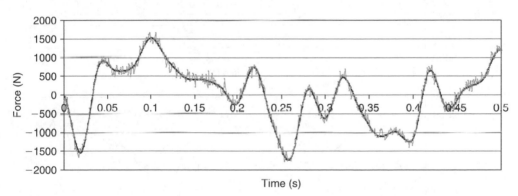

Figure 5.4 Example of identified force for 5-DOF under 10% noise Heavy line = actual force, light line = identified force

better option than the random force. To be able to identify the structures with limited output information only, and to achieve average error of less than 1% even under 10% noise is an accomplishment that, to the knowledge of the authors, has not been reported before. The feasibility of implementing the SSRM in eliminating the need for force measurement will also go a long way to developing strategies that will work on real systems.

In addition to the structural properties, the input force(s) is also identified. Figures 5.4 and 5.5 show examples of the forces identified for the 5 and 20-DOF

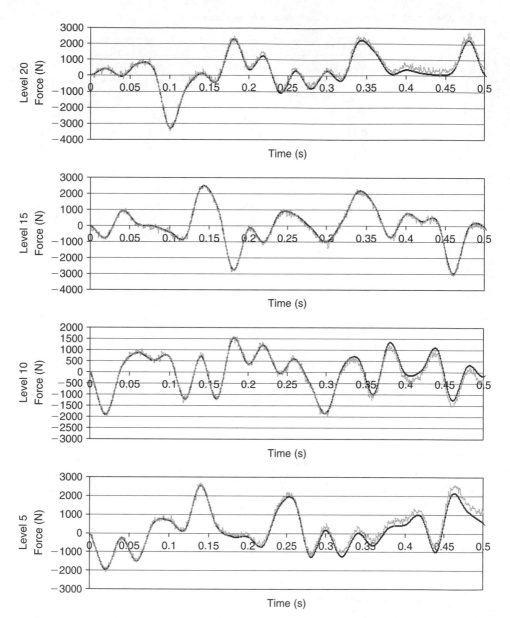

Figure 5.5 Example of identified forces for 20-DOF under 10% noise Heavy line = actual force, light line = identified force

structures respectively, compared to the actual input force used in the simulation. In both cases the identification result shown is the worst case of 500 data points and contamination with a very large 10% noise. The figures show that even under these conditions a very reasonable estimate of the force is achieved. For lower noise levels and longer data length, there is very little difference between the actual force and identified

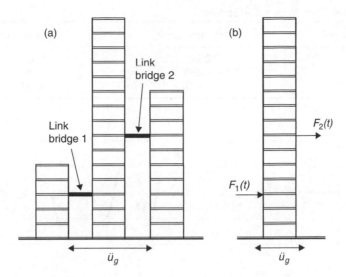

Figure 5.6   (a) Entire system of 3 connected structures, (b) structure extracted for identification

force; thus the comparison is not shown here. The high frequency variation observed in the figures is due to noise in the measured accelerations that is transferred to the force via the inertia term in the dynamic equilibrium equation (Eq. 5.3). This may be reduced by recalculating force based on updated accelerations. Nevertheless, since the identified stiffness values are very good, there is no need for this additional step.

5.3   Seismic Example

As a further example of how the output-only identification can be used, a seismically excited system of buildings is considered. The system consists of three shear buildings connected by two link bridges as illustrated in figure 5.6(a). The structural properties for the central building are $m_1-m_5 = 7 \times 10^5$ kg, $m_6-m_{15} = 4 \times 10^5$ kg, $k_1-k_5 = 6 \times 10^5$ kN/m and $k_6-k_{15} = 4 \times 10^5$ kN/m. The left and right buildings have $m = 4 \times 10^5$ kg and $k = 4 \times 10^5$ kN/m for all floors. The link bridges are modelled as linear springs with axial stiffness of 1×10^6 kN/m. The natural periods of the first two modes of the system are 1.6 s and 0.8 s. Damping is provided by Rayleigh damping with a damping ratio of 2% applied to the first 2 modes of the entire coupled system. For identification purposes α and β are identified along with the unknown stiffness and force. The mass of the building is assumed known and the seismic loading is easily computed from the measured ground motion. It should be noted here however that this is a case of input noise as the noise on the measured ground motions is passed directly to the excitation.

The response of the entire system to the first 5s of the NS component of the 1940 El Centro earthquake is simulated using a time step of 0.005s. The accelerations at the ground level and levels 2, 3, 4, 6, 7, 8, 10, 12, and 14 of the central building are extracted for use in the identification. Noise is added to all acceleration measurements

Table 5.4 Error in identified stiffness

Noise Level	Mean Error	Maximum Error
0%	0.34%	1.53%
5%	0.85%	2.08%
10%	1.47%	3.82%

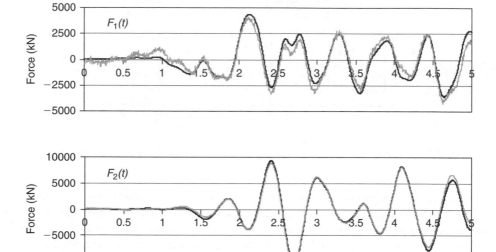

Figure 5.7 Example of forces identified under 10% noise (Heavy line = actual force, light line = identified force)

as white Gaussian noise scaled so that the RMS of the noise is a given percentage of the RMS of the response.

The link bridges cause coupling between the buildings which, when considering a single building, cannot be readily quantified. Using the proposed strategy, the central building and the coupling forces can be identified without any information of the other two buildings as shown in figure 5.6(b), where F_1 and F_2 represent the unknown coupling forces from the adjacent buildings.

A fairly broad search range, i.e. half to double the exact values, is again adopted for all unknown stiffness parameters. The reduced data length procedure is used in all cases with a reduced length of 40% of the data points used for 50% of the generations. The identification is considered in the presence of 0, 5 and 10% noise and for each case the analysis is repeated 5 times using fresh noise signals. In all cases the stiffness and damping parameters and the force(s) are identified. As our interest is in the identified stiffness values, the results presented in Table 5.4 compare the absolute mean and maximum errors in identified stiffness.

By identifying forces, rather than trying to measure them, we are able to avoid the error that would otherwise be passed through the simulation. The output-only results are very good; the achieved mean error is only 1.5% under 10% noise level.

In addition to identifying the structural properties, the method gives the time history of unknown input. Figure 5.7 shows an example of the forces identified compared to the actual force for a case of 10% noise contamination. The forces identified for 0% noise are not shown as they match the true force almost exactly. The forces identified under 10% noise are high pass filtered to remove the numerical drift that is inevitable for this sort of problem where only acceleration measurements are used. The figure shows that, even with limited and highly contaminated data, a very reasonable estimate of the force is achieved. As with the previous example, the high frequency variation observed in Figure 5.7 is due to noise in the measured accelerations that is transferred to the force via the inertia term in the dynamic equilibrium equation. This may be reduced by filtering or by recalculating force based on corrected accelerations. Nevertheless it is reckoned that the identification results are very good and this refinement step is not necessary as far as the main objective of identifying stiffness parameters is concerened. It is also important to note that the identification of F_2 is significantly better than that of F_1. This is most likely due to the fact that the second force is larger and acts at a location that causes a larger influence than the first force on the measured response of the structure.

5.4 Chapter Summary

An important extension of the structural identification strategy to output-only identification in the time domain has been described in this chapter. The strategy works by simultaneously computing the excitation forces as the structural parameters (stiffness and damping) are identified. The ability to identify structural properties in the time domain, without a need for force measurement, is not an entirely new concept. Nevertheless, to the authors' knowledge, the use of GA and the *simultaneous* computation of input force(s) is new, and the results achieved using the proposed strategy represent a big step forward. The ability to eliminate the need for force measurement allows time domain methods to compete more favourably with their frequency based counterparts and opens up the possibility for using a wider range of excitation methods, including natural vibrations such as wind, water or ground motions.

The strategy has been validated here using numerical simulations of forced and seismically excited structural systems. The results are outstanding, with mean errors in stiffness of less than 1% and maximum errors less than 3% achieved for structures with up to 20-DOF even when the incomplete acceleration measurements are contaminated with 10% noise. The very good results of identified stiffness is partly attributed to the improved GA strategy used and partly attributed to the avoidance of input measurement error which would otherwise reduce the identification accuracy. The output-only identification strategy will be put to test in the experimental verification study to be presented in Chapter 7.

Chapter 6

Structural Damage Detection

In this chapter, the SSRM strategy developed in the previous chapters is applied to the area of structural damage detection, an important area in structural health monitoring. In order to identify damage, the SSRM is used within a damage detection strategy as described in section 6.1. This strategy includes an option to fix the mass of the structure based on preliminary identification of the undamaged structure, and may also use the identified parameters to direct the search when identifying the damaged structure. An investigation into the effect of using these options is presented. The same three structures considered in the previous chapter are used and numerical trials are carried out in order to investigate how best to identify the damage. The knowledge gained is then used as the strategy is used to identify damaged members in a seven-storey steel model in the next chapter on experimental verification study.

6.1 Damage Detection Strategy

There are two possible scenarios when it comes to damage detection. (1) Damage can be identified with no prior measurement of the undamaged structure. (2) Damage can be identified utilising previous measurements. For the first scenario we have no choice but to identify the structural properties and compare these to some theoretical values in order to identify the magnitude and location of damage. In this case the SSRM can be utilised directly and no additional development is required. In principle, there are direct methods that can detect damage, for example, by means of fibre optic sensors to measure strains or detect cracks. Alternatively non-destructive test methods such as ultrasonic scanning can be used. These direct method are, however, not within the intended scope of this book which focuses on using vibration signals as a non-destructive and global means of damage detection. For the second scenario, however, the additional information of the undamaged structure can be utilised in developing an improved strategy. This section therefore deals with the scenario where we have measurements of the structure both before and after damage has taken place. The strategy assumes that the structural mass, stiffness and damping are unknown, and the damage can be quantified and detected as a change in the stiffness of the damaged member. Assuming the mass to be unknown allows the strategy to be applied to a wider range of problem and aids in proper calibration of the structural model. The

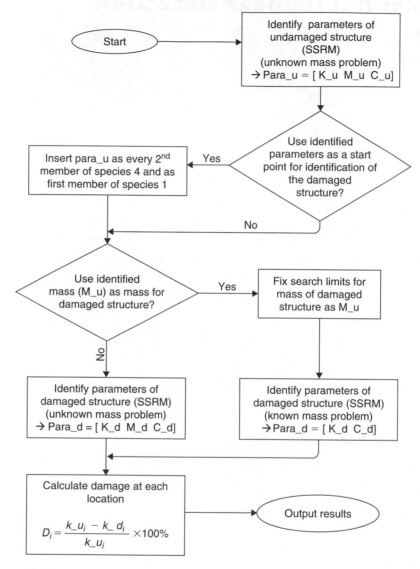

Figure 6.1 Damage detection strategy

aim is to detect the magnitude and location of the damage from the measured response of the structure before and after the damage takes place.

The proposed strategy is shown in figure 6.1. The strategy contains two additional options designed to improve identification to be conducted on the damaged structure. Firstly it is possible to use the parameters identified for the undamaged structure as a starting point for the identification of the damaged structure. This has obvious benefit in that only changes need to be identified giving the identification a good starting point resulting in a more accurate identification. This option is implemented by setting half of the individuals in species 4 to the values identified for the undamaged structure. The other half of the species as well as species 2 and 3 are initialized randomly. The value of half is chosen such that sufficient random results exist so as to ensure good performance of the crossover operation in the early stages.

Secondly there is an option of fixing the mass based on the mass of the undamaged structure. This option is useful if we are certain that the mass has not been altered since the measurement of the undamaged structure was made. The identification of the damaged structure is reduced from an unknown mass problem, to a much easier known mass problem, which can be identified with better speed and accuracy. There is also an added benefit that changes in stiffness will not be masked by an apparent change in the mass. For cases where significant changes in structural mass may have occurred this option should not be used and a full identification including the mass of the damaged structure must be done.

Finally the extent of the damage is calculated as the *loss in stiffness* of the structure as a percentage of the original undamaged stiffness. This damage measure is used as it better highlights the damaged members than integrity index type measures which express the remaining stiffness of the structure as a fraction of the original stiffness. Integrity index values of 0.99 and 0.95 may appear similar, whereas the corresponding damage extents of 1% and 5% are more clearly different and thus give a better feel for the damage that has occurred.

6.2 Verification of Strategy Using Simulated Data

In order to observe the performance of the damage detection strategy and to assess the effect of the options available, trials are carried using the same structures as chapter 5. In all cases the damage is simulated by a reduction of stiffness at the 4th storey. This storey represents a high, medium and low-level storey for the 5, 10 and 20-DOF systems, respectively. Damage magnitudes of 2.5%, 5% and 10% are simulated for each structure. In all cases the I/O noise level is set at 5%. All simulations are carried out for 500 data points at a time step of 0.01 s, with forces applied at every 5th level and acceleration measurements at 60% of levels as given for the unknown mass cases in table 6.2. Each system and damage level is simulated 25 times using newly generated random force and noise signals to produce 25 different data sets. In all cases the search limits are set, as before, to be half to double the actual parameters of the undamaged system.

The trials presented in this section are designed in order to observe the effect that the options of fixing the mass and using the undamaged parameters as a starting point will have on the identified damage. Therefore for each system and damage level there are four combinations of identification options to be tested. Only the identification of the damaged structure is affected by these options. The undamaged structure is treated as an unknown mass problem and the structural parameters are identified using the GA parameters given for unknown mass systems in table 6.1. If the mass is not fixed based on this result, the same unknown mass GA parameters are then used to identify the damaged structure. If the mass is fixed, the GA parameters for the known mass system can be used and the computational time greatly reduced. In all cases the reduced data length procedure is used with a reduced length of 200 used for 50% of the generations. The resulting computational times are indicated in table 6.1 for analysis conducted on a Pentium 4, 3-GHz PC. The total analysis time is the sum of the analysis for the

Table 6.1 Damage detection – GA Parameters

	Known Mass Systems			Unknown Mass Systems		
	5-DOF	10-DOF	20-DOF	5-DOF	10-DOF	20-DOF
Population size	7 × 3	9 × 3	19 × 3	45 × 3	65 × 3	90 × 3
Runs	4/10	4/10	4/10	5/15	5/15	5/15
Generations	60	100	200	150	200	300
Runs to ave for output	1	1	1	1	1	1
Crossover rate	0.8	0.8	0.8	0.4	0.4	0.4
Mutation rate	0.2	0.2	0.1	0.2	0.2	0.1
Window width	4.0	4.0	4.0	4.0	4.0	4.0
Migration	0.05	0.05	0.05	0.05	0.05	0.05
Regeneration	2	2	3	3	3	3
Reintroduction	25	30	50	50	100	150
Time (min:s)	0:07	0:30	4:10	3:00	10:40	44:30

Table 6.2 Damage detection of 5-DOF system

Use para_u	Fix mass	Identification Success %		Absolute Error in Damage			Maximum False Damage		
		2 ×	4 ×	Mean	Median	Max	Mean	Median	Max
2.5% damage									
Yes	Yes	100	84	0.11 (0.00)	0.08	0.46	0.30 (0.01)	0.23	1.10
Yes	No	100	68	0.41 (0.02)	0.31	1.67	0.55 (0.02)	0.37	1.97
No	Yes	96	72	0.23 (0.01)	0.14	1.07	0.50 (0.02)	0.45	2.47
No	No	72	40	1.07 (0.05)	0.62	4.31	1.19 (0.06)	0.70	6.78
5% damage									
Yes	Yes	100	96	0.21 (0.02)	0.09	2.06	0.34 (0.01)	0.26	1.31
Yes	No	96	84	0.52 (0.03)	0.30	3.20	0.79 (0.04)	0.44	3.45
No	Yes	100	100	0.18 (0.01)	0.09	0.57	0.43 (0.01)	0.36	0.98
No	No	92	76	0.94 (0.06)	0.56	7.78	1.32 (0.08)	0.84	9.60
10% damage									
Yes	Yes	100	100	0.13 (0.01)	0.08	0.51	0.52 (0.01)	0.51	1.60
Yes	No	100	100	0.55 (0.02)	0.35	1.99	0.80 (0.03)	0.50	2.13
No	Yes	100	100	0.25 (0.01)	0.17	1.12	0.59 (0.02)	0.40	1.47
No	No	100	96	1.08 (0.04)	0.85	3.27	0.73 (0.03)	0.46	3.94

* Noise level is 5% in all cases

undamaged and damaged structures depending on the option chosen. When the mass is fixed based on the undamaged parameters identified, the total times are 3 min 7 s, 11 min 10 s and 48 min 40 s for the 5, 10 and 20-DOF systems respectively.

A summary of the results is given in tables 6.2 to 6.4. There are three components to the results presented. These considerations can be understood by viewing the typical plot of damage results shown in figure 6.2, where the actual damage simulated is 2.5% in the 4th storey. The first component is the absolute error in the damage identified at the 4th (damaged) level. Just as important, however, is ensuring that damage is not falsely reported at undamaged levels. Thus the maximum false damage identified

Table 6.3 Damage detection of 10-DOF system

Use para_u	Fix mass	Identification Success %		Absolute Error in Damage			Maximum False Damage		
		2 ×	4 ×	Mean	Median	Max	Mean	Median	Max
2.5% damage									
Yes	Yes	88	80	0.22 (0.02)	0.08	0.85	0.49 (0.02)	0.29	2.31
Yes	No	72	48	0.65 (0.03)	0.31	3.04	1.41 (0.07)	0.88	7.85
No	Yes	80	68	0.24 (0.01)	0.12	1.64	0.84 (0.05)	0.32	4.67
No	No	44	40	0.99 (0.04)	0.61	3.67	2.19 (0.09)	1.99	7.34
5% damage									
Yes	Yes	96	80	0.21 (0.01)	0.12	1.15	0.63 (0.03)	0.36	2.88
Yes	No	76	60	0.55 (0.02)	0.48	1.92	1.91 (0.08)	0.98	7.36
No	Yes	88	76	0.22 (0.01)	0.16	0.64	0.90 (0.04)	0.36	4.20
No	No	84	68	1.00 (0.05)	0.44	5.02	1.19 (0.06)	0.45	6.28
10% damage									
Yes	Yes	100	96	0.13 (0.00)	0.09	0.39	0.45 (0.03)	0.22	3.89
Yes	No	96	92	0.68 (0.03)	0.44	2.20	1.35 (0.06)	0.79	6.18
No	Yes	100	100	0.26 (0.01)	0.16	1.23	0.67 (0.02)	0.47	1.96
No	No	92	76	1.18 (0.04)	0.91	5.10	1.84 (0.08)	1.33	6.93

* Noise level is 5% in all cases

Table 6.4 Damage detection of 20-DOF system

Use para_u	Fix mass	Identification Success %		Absolute Error In Damage			Maximum False Damage		
		2 ×	4 ×	Mean	Median	Max	Mean	Median	Max
2.5% damage									
Yes	Yes	88	76	0.11 (0.00)	0.07	0.39	0.55 (0.04)	0.19	4.04
Yes	No	84	56	0.28 (0.01)	0.20	0.95	0.78 (0.02)	0.50	2.66
No	Yes	92	80	0.24 (0.02)	0.09	2.23	0.58 (0.03)	0.35	2.96
No	No	72	40	0.63 (0.03)	0.45	2.70	1.34 (0.06)	0.74	5.44
5% damage									
Yes	Yes	92	84	0.14 (0.01)	0.08	0.82	0.57 (0.03)	0.20	2.93
Yes	No	92	76	0.34 (0.02)	0.20	2.26	1.04 (0.05)	0.56	5.24
No	Yes	92	88	0.21 (0.01)	0.10	0.93	0.74 (0.04)	0.36	4.97
No	No	88	76	0.58 (0.02)	0.28	1.97	1.32 (0.05)	0.84	5.74
10% damage									
Yes	Yes	100	100	0.14 (0.01)	0.09	0.79	0.42 (0.02)	0.29	1.96
Yes	No	100	96	0.68 (0.06)	0.29	6.47	1.20 (0.03)	1.05	3.59
No	Yes	100	96	0.15 (0.01)	0.12	0.46	0.74 (0.02)	0.52	2.96
No	No	100	88	0.61 (0.05)	0.22	5.36	1.24 (0.05)	0.84	4.51

* Noise level is 5% in all cases

on the undamaged floors is also presented. For both of these error considerations the mean, median and maximum values over the 25 tests are presented. In many cases it is seen that one bad result distorts the mean, and the median may give a better indication of the expected performance. As with the results of the previous chapter, the standard errors of the mean results are given in brackets next to the mean values. A graphical

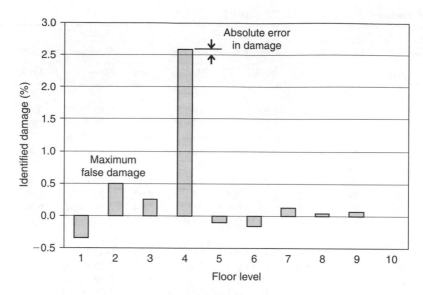

Figure 6.2 Example identification result for 10-DOF system

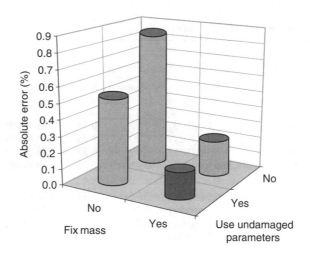

Figure 6.3a Mean identification results for absolute error in damage

overview of the mean results obtained over all systems and all damage scenarios is presented in figure 6.3.

Finally, to be of practical use the damage identified should exceed any false damage by a reasonable margin. In this regard, the *success* of identification is herein defined as, among the tests done, the proportion where the identified true damage exceeds the maximum false damage by a given ratio (which is 2 or 4 as shown in the Tables). For example, the per cent shown under "Success" for "4X" is the success rate corresponding to tests where the true damage identified at the damaged level is at least 4 times the maximum false damage identified at any of the undamaged levels.

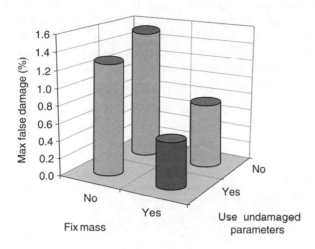

Figure 6.3b Mean identification results for maximum false damage

The results demonstrate that the strategy is able to accurately and consistently identify even small levels of damage corresponding to a change in storey stiffness of only 2.5%. The performance of the strategy is enhanced by using the option of fixing the mass based on the result of the undamaged structure. The results also improve, although to a lesser extent, when the undamaged parameters identified are used as a starting point for the identification of the damaged structure. In addition to improved accuracy, fixing the mass based on the undamaged result reduces the computational time significantly. For a 20-DOF system the time taken to identify the damaged structure is reduced from 44 min 30 s to only 4 min 10 s. The results also highlight an important fact that although the identification is very good, it is not perfect. In some cases the identification will fail from the standpoint that identified damage should be significantly larger than any false damage. This is seen for example in the results for the 10-DOF system with 2.5% damage at the 4th level. In three cases (12%) of the 25 trials conducted, the damage identified at the 4th level was not more than twice that of any other level in the structure. In one of those cases the damage identified at an undamaged floor actually exceeded that identified at the damaged (4th) level. While this failure only occurs once in 25 trials conducted it should be considered. In an analysis of a real structure it would therefore be recommended that the identification test be carried out more than once to ensure consistency and validity of the result as it would be highly unlikely that the same false result would be identified more than once.

6.3 Chapter Summary

This chapter has introduced a damage detection strategy utilising the SSRM developed in the previous chapters. The strategy makes use of measurement of both the undamaged and damaged structures to significantly improve the accuracy and reliability of detection. The strategy uses measurement of the undamaged structure in order to calibrate the structural model, and by fixing the mass of the structure based on the identification of the undamaged structure, is able to reduce the damage detection step to a simpler known mass problem. The parameters identified during the calibration step are also used to initiate the search when identifying the damaged structure. The numerical studies presented demonstrate that small levels of damage representing as low as 2.5% reduction in stiffness can be accurately and consistently identified in the presence of 5% I/O noise. In the next chapter this strategy will be further validated experimentally by damage detection of a 7-storey steel model.

Chapter 7

Experimental Verification Study

In order to assess the effectiveness of the strategy on more realistic data, a seven-storey steel model is constructed and tested in the laboratory. The model layout and dimensions are given in figure 7.1. The structure is designed with flexible columns (provided by thin plate members) and relatively rigid beams (constructed from square hollow section) to provide a shear building behaviour. Combined with the symmetry of the structure and loading, this reduces the significant motion to a single translation at each floor level. The mass of the structure is also concentrated at the floor levels meaning the lumped mass formulation should prove reasonable. For reference purpose, the levels are labelled from 1 to 7, with 1 being the bottom level and 7 the seventh (roof) level.

As a basis for comparison of results, and to help plan the identification tests, preliminary calculations and testing is first carried out as presented in section 7.1. This includes calculations of the estimated mass and stiffness matrices assuming a steel modulus of 205 GPa and density of 7850 kg/m^3. The section also includes static tests to estimate the as-built stiffness of the structure, as well as some dynamic tests to establish the as-built natural frequencies.

Following these initial tests on the model, the testing procedure for the main identification tests is described in section 7.2. This outlines the tests that are to be conducted and describes the damaged scenarios to be identified. Finally, the analysis of the identification tests is presented in section 7.3.

7.1 Preliminary Calculations and Testing

7.1.1 Estimation of Structural Properties

Mass
The mass of each floor is estimated by lumping the distributed mass of the structure at the nearest floor level. The mass is calculated based on the member sizes shown in figure 5.4 and using a mass density of 7850 kg/m^3. The mass of welds is ignored and results rounded off to the nearest 10 g. The mass matrix (diagonal) of the structure is therefore;

$$\mathbf{M} = diag(3.78 \quad 3.78 \quad 3.78 \quad 3.78 \quad 3.78 \quad 3.78 \quad 3.31)\,kg$$

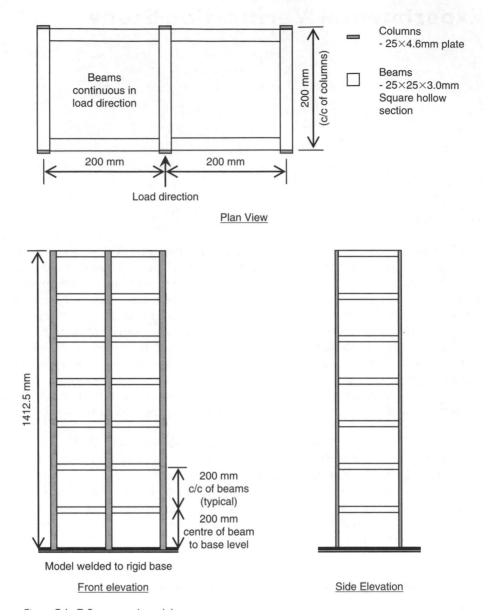

Figure 7.1   7-Story steel model

Stiffness

The stiffness is estimated by slope-deflection considerations assuming pure horizontal translation of the floors. The columns are assumed to start from the mid height of the beam rather than the beam face to account for the fact that some small rotation at the beam face may occur. The stiffness is then estimated as given below.

$$k = \frac{12EI}{L^3} = 62.36 \text{ kN/m (per column)} = 375 \text{ kN/m (per level)}$$

Table 7.1 Calculated natural frequencies

	ω (rad/s)	f (Hz)	T (s)
Mode 1	66.95	10.65	0.094
Mode 2	197.69	31.46	0.032
Mode 3	319.16	50.80	0.020
Mode 4	425.83	67.77	0.015
Mode 5	513.06	81.66	0.012
Mode 6	577.42	91.90	0.011
Mode 7	616.74	98.16	0.010

where L = centre-to-centre length

$$\mathbf{K} = \begin{bmatrix} 750 & -375 & & & & & 0 \\ -375 & 750 & -375 & & & & \\ & -375 & 750 & -375 & & & \\ & & -375 & 750 & -375 & & \\ & & & -375 & 750 & -375 & \\ & & & & -375 & 750 & -375 \\ 0 & & & & & -375 & 375 \end{bmatrix} \times 10^3 \, \text{N/m}$$

In comparison if the columns are assumed fixed at the beam face, the storey stiffness is 560 kN/m.

Natural Frequencies

The natural frequencies of the system are obtained by solving the eigenvalue problem as given in equation 7.1 and the resulting frequencies are displayed in table 7.1.

$$|\mathbf{K} - \omega^2 \mathbf{M}| = 0 \tag{7.1}$$

7.1.2 Static Tests

In order to get an estimate of the as-built stiffness of the structure, static tests are performed. The model was mounted horizontally to a rigid vertical support as indicated in figure 7.2. This allowed weights to be hung from the floors while displacement transducers recorded the displacement. Two displacement transducers were used (one on each side of the model) and the average displacement taken. The difference between the two displacements also allows us to observe any rotational coupling that may be present. Several different weights were used and the stiffness determined from the slope of the regression line plotted through the load-displacement points obtained. The procedure was repeated for each level starting from the first level and working outwards. The test determines the total stiffness of the structure up to the given floor and thus the stiffness of the individual floors must be calculated from these

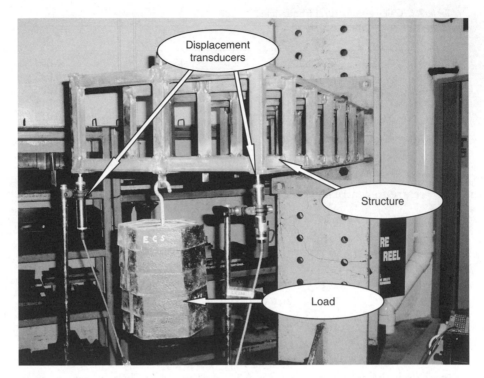

Figure 7.2 Static test

results as;

$$K_1 = K_{T1}$$

$$\frac{1}{K_2} = \frac{1}{K_{T2}} - \frac{1}{K_1} = \frac{1}{K_{T2}} - \frac{1}{K_{T1}}$$

$$\frac{1}{K_3} = \frac{1}{K_{T3}} - \frac{1}{K_2} - \frac{1}{K_1} = \frac{1}{K_{T3}} - \frac{1}{K_{T2}} \qquad (7.2)$$

$$or \quad K_i = \frac{1}{\left(\frac{1}{K_{Ti}} - \frac{1}{K_{T(i-1)}} \right)}$$

where K_i is the storey stiffness and K_{Ti} represents the total stiffness of the structure determined by the slope of the regression of the displacements at level i due to load applied at level i.

Based on the measurements in the static tests, the stiffness values are obtained and presented in table 7.2. For comparison, the stiffness calculated in the previous section (based on centre-to-centre lengths) is also displayed in the table. It is seen that the as-built structure is actually slightly stiffer than the theoretical values. Back calculation for $K = 450$ kN/m and cross section of 25×4.6 mm, gives an effective column length (fixed ends) of 188 mm showing the result to be reasonable if the column to beam

Table 7.2 Static stiffness of model

	Calculated (kN/m) based on centre-to-centre lengths	Experimental for the as-built frame (kN/m)
K_1	375	409.97
K_2	375	505.43
K_3	375	452.63
K_4	375	482.44
K_5	375	411.03
K_6	375	478.86
K_7	375	403.15

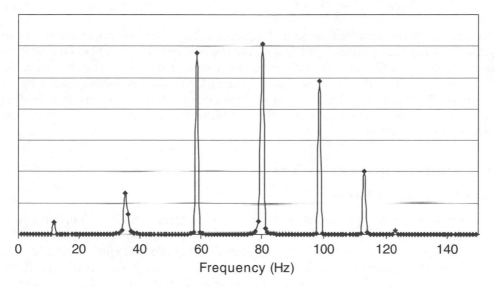

Frequency (Hz)

Figure 7.3 Power spectrum of response at level 7 due to impact at level 7

connection is good. The difference could also be due to slight variation in the member cross section. Variation in the thickness of only 0.1 mm would cause a 7% change in stiffness. These measured values are considered more accurate than the calculated values and should form the basis for comparison with identification values.

7.1.3 Impact Tests

In order to determine the natural frequencies of the structure, impact tests were carried out. The structure was excited using a hammer and the response measured with accelerometers at each floor and recorded using a 16 channel digital oscilloscope at a sampling rate of 2 kS/s. The algorithm of fast Fourier transform (FFT) was then used to convert the signal to frequency domain and the structural frequencies observed from the plot of power spectrum. 2048 data points were used for the FFT resulting in 1024 frequency divisions. The frequencies obtained are therefore accurate to approximately ±0.5 Hz. An example of the plot obtained is shown in figure 7.3 for the case of an impact at level 7 and measurement also at level 7. The response at other floors and

Table 7.3 As-built structural frequencies

	Calculated (Hz)[#]	Measured (Hz)	Measured Period (s)
Mode 1	11.7	11.7	0.085
Mode 2	34.0	35.2	0.028
Mode 3	55.3	58.6	0.017
Mode 4	72.3	80.1	0.012
Mode 5	90.0	98.6	0.010
Mode 6	101.4	113.3	0.009
Mode 7	108.6	123.0	0.008

[#] Frequencies calculated using the theoretical masses and as built stiffness values from table 7.2.

for other impacts identified the same frequencies. The extracted values are shown in table 7.3 where comparison is made with the frequencies calculated based on the stiffness values obtained from the static tests. It is seen that the results match reasonably well, particularly in the first few modes. The FFT also showed significant energy in the 600–1000 Hz range. This is most probably due to local vibrations of individual columns. Care must be taken in planning the identification experiments so that these modes are not excited.

7.2 Main Identification Tests

7.2.1 Excitation Force

For many of the tests presented in the previous chapters random signals were used as excitation. Nevertheless it is more repeatable to use a smoother waveform. The excitation forces used for the tests must balance having broad frequency content with being smooth enough that they can be accurately integrated for generation of the response in the identification procedure. Similar to the tests presented for the output-only identification, in order to create a smoothed random force for input into the function generator, a random force is first generated at a time step of 0.004 s (250 Hz) and the signal is then converted to a time step of 0.0002 s (5000 Hz) by interpolating and smoothing the force over the intermediate data points. The same interpolation procedure was used here and five different input forces were prepared for use in the tests. The input forces, labelled A to E were prepared for 500 data points, representing a 0.1 s time interval and are illustrated in figure 7.4.

7.2.2 Test Setup and Procedure

A schematic diagram of the dynamic testing and data acquisition system is shown in figure 7.5 while figure 7.6 shows the system as it was used in the lab. The forces described in the previous section were input into the signal generator (Signametrics function/pulse generator, model SM-1020) in the PC as a *.wav file. The signal was then passed through a power amplifier in order to produce sufficient power for the electromagnetic shaker. The force generated by the shaker was transferred to the structure via a connecting rod at the 7th storey and the force measured by an ICP (Integrated

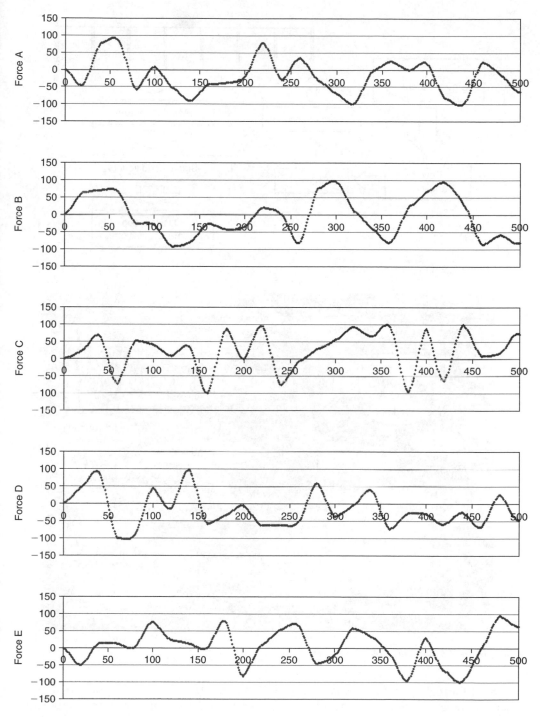

Figure 7.4 Input forces

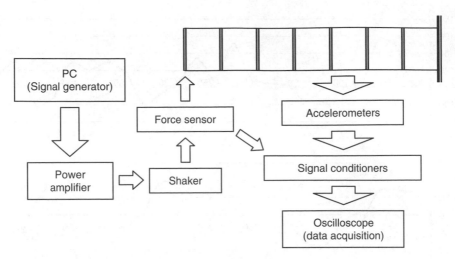

Figure 7.5 Diagram of test setup

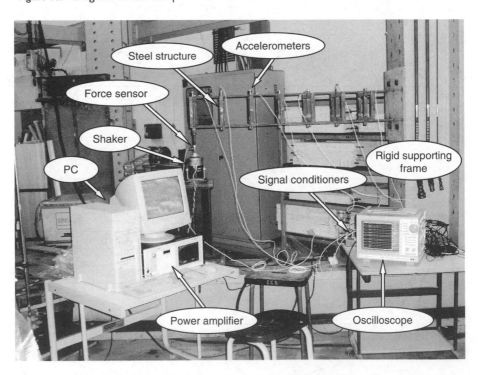

Figure 7.6 Test setup used in the lab

Circuit Piezoelectric) force sensor (model PCB-208C02). Figure 7.7 shows the shaker-sensor-structure connection detail. The shaker was rigidly mounted to the supporting frame using a bolted connection. The force sensor was connected to the shaker by a threaded stainless steel stringer. The sensor was then attached to an aluminium base

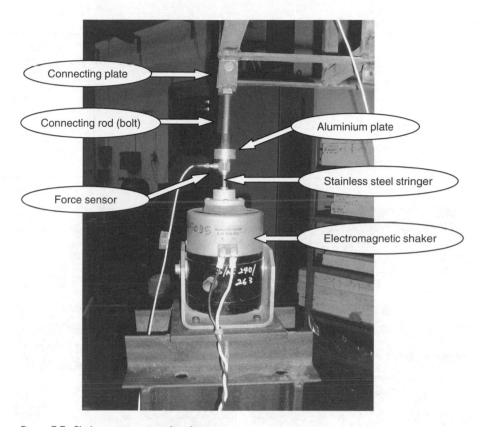

Figure 7.7 Shaker connection detail

plate which is threaded to accommodate a standard bolt. A small connecting plate was welded to the top of the bolt to fix the assembly to the structure.

The response of the structure was measured using seven ICP accelerometers mounted at the top of each level. The accelerometers used are described in table 7.4 they were attached to the structure where possible by threaded connections to nuts mounted on the structure using epoxy as shown in figure 7.8a. Where a threaded connection was not available, thin double sided tape was used as shown in figure 7.8b. The signals from the force sensor and the accelerometers are passed through signal conditioners and recorded using a 16-channel digital oscilloscope. The data was recorded on the oscilloscope at a sampling rate of 5 kS/s. Although the highest frequency of interest (the 7th mode) is only 123 Hz, this high sampling frequency allows for a better capture of the excitation facilitating a more accurate simulation of the response during identification. 10,000 points were recorded during testing and then 500 points starting from just before the application of the force are extracted and copied to the input file for the damage detection program.

Before being used, the data was processed by removing any mean offset that might have existed. For this purpose a sample of 500 data points (representing 0.1 s) immediately preceding the application of the force was used. It should be noted that this is

Table 7.4 Accelerometer specification

Level	Model	Serial No	Range	Sensitivity (mV/g)	Freq. Range (Hz)
1	PCB – 312A	6532	±50g	92.6	1–2000
2	PCB – 302A	17618	±500g	10.02	1–3000
3	PCB – 353B	83737	±50g	99.1	1–4000
4	PCB – 312A	6533	±50g	88.6	1–2000
5	PCB – 308B	31764	±50g	100	1–3000
6	PCB – 302A	14185	±500g	9.99	1–3000
7	PCB – 308B	31765	±50g	99.6	1–3000

Figure 7.8 Mounting of accelerometers: (a) threaded, (b) double sided tape

the only signal processing used in this study. The noise level in the signal can also be estimated from this pre-event portion of the record by comparing the standard deviation of the pre-event portion to that of the 500 points used in the identification. For the tests conducted, the noise level ranged from 1–10% on all signals.

7.2.3   *Damage Scenarios*

Two damage magnitudes and three damage locations are used in order to examine the performance of the strategy. The damage magnitudes are classified as small and large while the location is given by the corresponding level number. The undamaged structure and six damaged scenarios are considered as described in table 7.5. The approximate magnitude of the damage is 4.1% for small damage and 16.7% for large damage as discussed further below. In addition to these six basic cases, more damage scenarios can be considered by treating one of the damaged conditions as the undamaged structure. For example, if damage scenario 2 is treated as the undamaged structure, then the additional damage of damage scenario 3 is that of a small damage at level 6. In this way, without conducting further experimental tests, additional cases of small damage

Table 7.5 Basic damage scenarios

	Small Damage	Large Damage
D0	–	–
D1	Level 4	–
D2	–	Level 4
D3	Level 6	Level 4
D4	Levels 3 and 6	Level 4
D5	Level 3	Levels 4 and 6
D6	–	Levels 3, 4 and 6

Table 7.6 Additional damage scenarios

	Undamaged Case	Damaged Case	Resulting Small Damage
D7	D2	D3	Level 6
D8	D3	D4	Level 3
D9	D2	D4	Levels 3 and 6

at one or more levels are generated as shown in table 7.6 to test the GA-based damage identification strategy.

Application of Damage

Controlled damage is created by cutting the members at the proposed levels. In all cases the centre column at the lower side of the structure is cut to avoid disturbing the accelerometers attached to the top of the structure. Small damage is formed as partial cuts near the top and bottom of the column as indicated in figure 7.9, whereas large damage is created by a complete cut at the mid-height of the column. The cuts for small damage are placed near to the beam column connection in order to be in an area of high bending. An example of the cuts applied to achieve small and large damage is shown in figure 7.10.

The expected reduction in stiffness due to the small cuts is estimated by the finite element analysis (ABAQUS 1998). As cuts are made at both the top and bottom of the column, the bending remains symmetric about the mid-height of the column and only half the column needs to be modelled, with one end fixed and the other free due to the inflection point that exists at the mid-height position. The FEM model of both the damaged and undamaged column is shown in figure 7.11. Shell elements are used with a grid size of 2.5 mm. An arbitrary load of 100 N is applied as a series of nodal loads along the free edge. The cut is simulated by the removal of 3 elements on each side resulting in 7.5 mm long, 2.5 mm wide cuts. The resulting displacements are noted and compared in table 7.7 to determine the change in stiffness. The analysis is repeated for a mesh size of 1.25 mm in order to observe the effect of the modelling on the stiffness obtained. In this case the cut is modelled as 1.25 mm wide. As there are 6 columns per floor, the expected reduction in column stiffness of 24.6% will result in a reduction

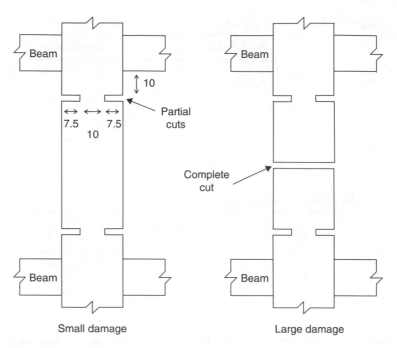

Figure 7.9 Illustration of damage

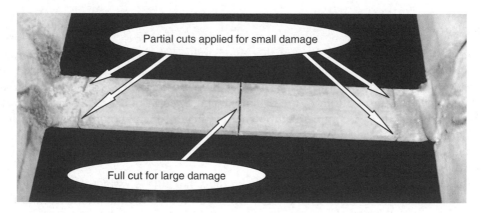

Figure 7.10 Damage applied to the structure

in storey stiffness of only 4.1%. For the case of large damage the column damage is 100% resulting in a storey stiffness reduction of 16.7%.

7.3 Experimental Identification Results

7.3.1 Identification of Undamaged Structure

Identification of the undamaged structure is first carried out in order to observe the variation in structural parameters from those predicted or obtained from the static

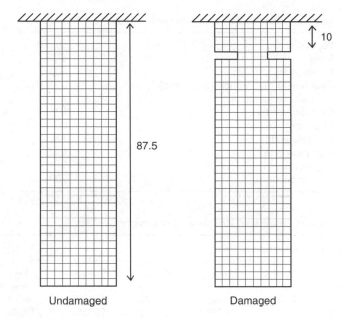

Undamaged   Damaged

Figure 7.11 FEM model for small damage

Table 7.7 Result of FEM analysis for small damage

Column	Mesh	Displacement	Column Stiffness	Column Damage
Undamaged	2.5 mm	0.527 mm	94.88	–
Undamaged	1.25 mm	0.527 mm	94.88	–
Damaged	2.5 mm	0.713 mm	70.13	26.1%
Damaged	1.25 mm	0.699 mm	71.53	24.6 %

tests. The GA parameters used for the identification are given in table 7.8 and the search limits are set as 150 to 800 kN/m for stiffness, 2 to 5.5 kg for mass, 0 to 4 for α and 0 to 0.0002 for β. The computational time required is 30 min 30 s for a Pentium 4, 3-GHz PC.

A summary of the results obtained is presented in table 7.9 and figures 7.12 and 7.13. These results highlight the fact that the dynamic model does not perfectly represent the structural system, and that the first stage in the damage detection strategy serves to 'calibrate' the model so damage can be better identified. Because of this, the identification is more consistent for tests using the same force input, but identified parameters may vary more if different forces are is used. This fact is highlighted in the damage detection results in the following sections. In general, the identified stiffness of the structure is higher than the stiffness identified in the static tests earlier. The exception is the first story, where the stiffness is significantly lower. This may be due to a less rigid connection at the base of the structure. The mass is generally lower than the calculated values, most probably due to variation in member thickness and modelling error in the

Table 7.8 Identification of undamaged structure – GA parameters

Population Size	50 × 3
Runs	5/20
Generations	800
Data length (full/reduced/% of time)	500/200/50
Runs to ave for output	1
Crossover rate	0.4
Mutation rate	0.2
Window width	4.0
Migration rate	0.05
Regenerations	3
Reintroductions	200

Table 7.9 Identification of undamaged structure – Dynamic test results

	Mean identification results						
	Force A	Force B	Force C	Force D	Force E	Average	Range (%)*
K1	295411	314836	262821	312228	283598	293779	17.71
K2	532176	548887	503918	553384	507057	529084	9.35
K3	506414	525520	504960	498751	510883	509306	5.26
K4	482383	470840	486672	478544	495053	482698	5.02
K5	504470	481815	501969	499633	489848	495547	4.57
K6	504667	533153	512566	496872	510645	511581	7.09
K7	491685	503471	502893	503418	491830	498659	2.36
M1	3.449	3.378	3.211	3.352	3.269	3.332	7.14
M2	3.189	3.329	3.159	3.350	3.239	3.253	5.87
M3	3.370	3.402	3.381	3.517	3.311	3.396	6.07
M4	3.378	3.617	3.454	3.428	3.442	3.464	6.90
M5	3.328	3.396	3.413	3.389	3.387	3.383	2.51
M6	3.264	3.458	3.258	3.326	3.160	3.293	9.05
M7	3.473	3.567	3.400	3.470	3.522	3.486	4.79
Alpha	0.202451	1.078361	0.086371	0.295406	0.000011	0.332520	324.30
Beta	0.000000	0.000000	0.000000	0.000000	0.000000	0.000000	–
Frequencies#							
Mode 1	11.85	11.82	11.51	11.90	11.71	11.76	3.32
Mode 2	35.42	35.31	35.20	35.31	35.29	35.30	0.61
Mode 3	58.17	58.34	58.24	58.18	58.23	58.23	0.28
Mode 4	79.62	79.57	79.63	79.63	79.60	79.61	0.07
Mode 5	98.15	96.60	98.13	98.08	98.10	97.81	1.59
Mode 6	113.41	113.22	113.30	112.46	113.34	113.15	0.84
Mode 7	120.60	119.20	120.09	118.85	120.44	119.84	1.47

*The range is the difference between the maximum and minimum identified value, expressed as a percentage of the average.
#Frequencies calculated from solving the eigenvalues from the identified stiffness and mass.

assumption of mass as lumped values. The overestimation of mass at the 7th storey is reasonable as part of the shaker connection is included in the identified mass. It is also interesting to observe the frequencies indirectly identified by the strategy. The identification of these frequencies is considered indirect as they are not specifically

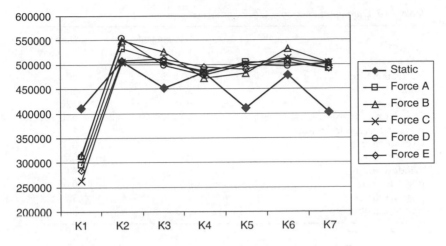

Figure 7.12 Dynamic tests – Identification of undamaged structure, stiffness

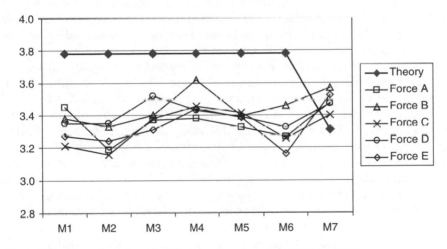

Figure 7.13 Dynamic tests – Identification of undamaged structure, mass

identified, but can be calculated from the identified mass and stiffness properties. It is noted that, while the identified parameters vary by more than 5%, the frequencies vary only very slightly. This observation is consistent with the earlier discussion in chapter 1, where it was mentioned that natural frequencies tend to be relatively insensitive to local changes in structural properties. In all cases the identified frequencies match the measured frequencies from table 7.3 with very good accuracy. The identified damping shows significant variation, indicating the assumed damping model may be inappropriate. As the damping of the structure is reasonably small and only a short time-history is required, this damping is unlikely to have a significant effect on the main objective of identifying stiffness.

Table 7.10 Damage detection – GA Parameters

	Undamaged Structure	Damaged Structure
Population Size	50 × 3	20 × 3
Runs	5/20	5/15
Generations	800	120
Data length (full/reduced/% of gen)	500/200/50	500/200/50
Runs to average for output	1	1
Crossover rate	0.4	0.8
Mutation rate	0.2	0.2
Window width	4.0	4.0
Migration rate	0.05	0.05
Regenerations	3	3
Reintroductions	200	30

7.3.2 Damage Detection

The damage detection tests are carried out using the GA parameters given in table 7.10. For the identification of the damaged structure the mass is fixed based on the result of the undamaged structure. The stiffness and damping parameters identified for the undamaged structure are also used as the starting point for identification of the damaged structure as was described for the earlier numerical examples. Using the GA parameters given, the computational time is 30 min 30 s for the identification of the undamaged structure (unknown mass) and computer time of 1 min 25 s for the damaged structure (known mass) on a Pentium 4, 3-GHz PC, resulting in a total analysis time of 32 minutes. When considering this computational time it is important to remember the identification of the undamaged structure is in effect a calibration step and only needs to be carried out once on the undamaged structure. Subsequently the identification of damage then requires 1min 25 s which is very fast from a practical point of view and can be performed on site if necessary to check for damage. As in the previous section, the search limits are set as 150 to 800 kN/m for stiffness, 2 to 5.5 kg for mass, 0 to 4 for α and 0 to 0.0002 for β.

The results of the damage detection for the steel model are discussed in two parts. Firstly, the effect of the input forces used for the undamaged and damaged structures is examined. For this purpose, full measurement of the structure is used and all combinations of the five forces are considered. Following this the effect of incomplete measurement is investigated by carrying out the identification using 4 and then only two acceleration measurements. In these tests, only the case of the same input force for the undamaged and damaged structures is considered.

Input Force

The tests carried out using full measurement are separated into two groups depending on whether the force used to identify the damaged structure is the same as, or differs from, that used to identify the undamaged structure. That is, if force A is used to identify the undamaged structure (calibration), force A is again used to identify the

Table 7.11 Damage detection results based on same force input for undamaged and damaged structure and full measurement

Damage Scenario		Mean damage identified (std dev)			Success (%)		
		Small	Large	Max False Damage	1X	2X	4X
Undamaged	D0			0.72 (0.24)	–	–	–
Single small	D1	3.54 (0.36)		1.43 (0.78)	93%	67%	33%
	D7	4.56 (0.46)		1.21 (0.66)	100%	89%	62%
	D8	4.16 (0.78)		1.08 (0.44)	100%	87%	49%
					98%	81%	48%
Two small	D9	3.90 (0.79)		1.05 (0.34)	96%	87%	51%
		4.45 (0.43)					
Single large	D2		17.19 (1.97)	4.06 (1.61)	100%	100%	47%
Three large	D6		18.94 (3.06)	4.29 (1.76)	100%	91%	53%
			16.01 (1.84)				
			18.99 (1.10)				
One small and one large	D3	4.11 (0.47)	17.55 (2.03)	4.19 (1.37)	47%	9%	0%
Two small and one large	D4	4.46 (2.10)	17.29 (2.00)	4.43 (1.93)	31%	13%	9%
		4.39 (1.93)					
One small and two large	D5	5.34 (2.58)	17.29 (1.31)	2.44 (1.96)	87%	40%	27%
			19.71 (1.59)				

damaged structure, or a different force (say force B) is used. A summary of the results are given in tables 7.11 and 7.12.

It is important that the strategy does not report the structure as damaged when it is in fact not damaged. The discussion here therefore first considers the identification results using the tests on the undamaged structure for both undamaged and damaged inputs (damage case D0). When the same input force is used, the maximum false damage reported averages only 0.72%. In all of the 45 combinations tested the maximum false damage identified is 2.32% in the worst case, and only 4 of the 45 tests (9%) exceed 2% maximum false damage, indicating a very good result. Where different input forces are used, the maximum false damage averages 5.45% and is above 4% in 53% of the 180 cases considered. These errors suggest that identification of damage using the same input forces will be significantly more reliable than cases where different forces are used.

The results summarised in table 7.11 give the identification of the damaged cases achieved when the input forces are the same. It is seen that the results are excellent for cases where a single magnitude of damage is to be detected. The magnitude and location of damage is accurately identified and the ratio of damage to maximum false damage (success) is very good. The identification is less satisfactory when multiple damages of *different magnitudes* are to be identified. In these cases the large damage (17%) present often causes false damage to be reported at other levels. When the false

Table 7.12 Damage detection results based on different force input for undamaged and damaged structure and full measurement

Damage Scenario		Mean damage identified (std dev)			Success (%)		
		Small	Large	Max False Damage	1X	2X	4X
Undamaged	D0			5.45 (3.32)	–	–	–
Single small	D1	2.58 (4.28)		5.93 (3.97)	50%	22%	2%
	D7	4.68 (2.01)		5.78 (2.86)	39%	12%	0%
	D8	3.43 (4.05)		5.29 (3.25)	51%	23%	3%
					47%	19%	2%
Two small	D9	3.27 (4.19)		5.68 (3.33)	26%	7%	4%
		4.53 (2.26)					
Single large	D2		16.84 (3.72)	6.53 (4.91)	83%	70%	54%
Three large	D6		19.19 (3.81)	5.04 (4.62)	86%	80%	53%
			16.10 (2.87)				
			19.64 (2.12)				
One small and one large	D3	4.42 (1.72)	17.23 (3.13)	6.37 (5.12)	51%	24%	7%
Two small and one large	D4	5.21 (3.45)	16.94 (3.37)	5.90 (5.76)	49%	31%	17%
		4.39 (1.85)					
One small and two large	D5	5.62 (3.83)	16.84 (2.89)	4.17 (4.03)	63%	56%	30%
			20.22 (2.16)				

damage is in the order of 4%, identification of small real damage of 4% becomes impossible. It should be noted it is only the small damage that is unable to be correctly identified in these cases. For practical purposes these cases should be considered as a partial success as the large (more important) damage is still successfully identified in almost all cases. The success percentages given in the table are low as in this study success requires that *all* damage levels be properly identified. For the cases where there is a single or multiple damage of *similar magnitude* the results are excellent and the damaged level is successfully reported as containing the largest damage in more than 95% of cases, even when the damage is only 4%. In more than 80% of these cases the damage identified is more than double that of any false damage reported.

The typical results illustrated in figure 7.14 help to illustrate the above discussion. It is seen that the single small damage of D1 and the single large damage of D2 are clearly identified, as is the multiple small damage of D9. However for damage case 3, while the large damage at level 4 and the small damage at level 6 are both correctly identified, the false damage reported for level 1 could mask the identification of the real damage at level 6. In this case the large damage at level 4 would be properly identified, but it would be difficult to distinguish the real small damage from the false damage.

Due to modelling imperfections, the identification of the undamaged structure should be thought of as calibration to reduce modelling error. This fact must be kept in mind when considering the results summarised in table 7.12, where the force used to identify the damaged structure differs from that used to identify the undamaged

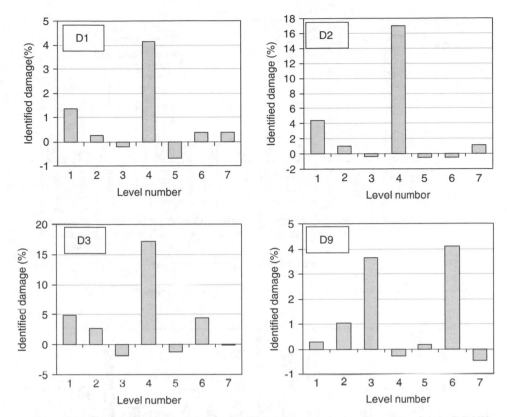

Figure 7.14 Typical identification results for full measurement using same input forces D1 (4% at level 4), D2 (17% at level 4), D3 (17% at level 4 and 4% at level 6), D9 (4% at levels 3 and 6)

structure. In this case the identification success is significantly reduced as the model is calibrated using one force and then used to detect the damage using another force. For the case of large damage the effect of modelling error can be overcome, and identification is reasonably good. For the small damage cases, errors in modelling could be larger than the real damage and the damage detection is unreliable, due to significant damage falsely identified at undamaged floors.

Figures 7.15 and 7.16 clearly highlight this fact with a comparison of the results using same and different forces. The accuracy of the magnitude of identified damage is of course important. It is seen from tables 7.11 and 7.12 that on average the damage magnitude identified is reasonably good; however the standard deviation is very large in some cases. The ratio of standard deviation and mean (coefficient of variation) is used to compare the results in figure 7.15. It is seen that the results vary much more for the cases where different input forces are used, particularly for the case of small damage where the variation is very large. This variation is one of the reasons for the lower success rate observed in figure 7.16. The other reason, as mentioned earlier is that due to modelling error, the maximum false damage is much larger when identification is carried out using different forces.

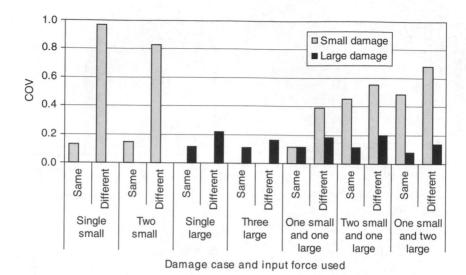

Figure 7.15 Effect of input force on identification – Variation of identified damage

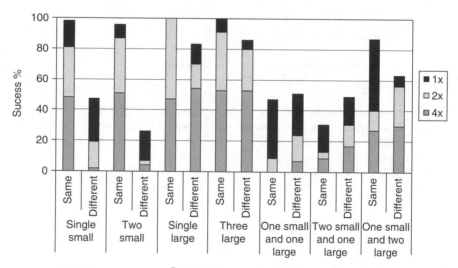

Figure 7.16 Effect of input force on identification – Success %

Incomplete Measurement

Identification of the structure using incomplete measurement is important for the extension of the method into more realistic systems. The measurements taken previously are used, but the available data are reduced to only include selected acceleration measurements. The measurement is first reduced to the four odd numbered levels in the structure and then to only two levels, namely, levels 2 and 6.

As with the case of full measurement we first consider the ability of the strategy to detect the undamaged structure. The maximum false damage averages 1.24% and 0.96% for the case of four and two measurements respectively. The success in terms of the number of results exceeding 2% is 24% and 11% respectively, and in only one instance the maximum false damage exceeded 4%. The results in both cases are very good and give confidence that the identification of the damaged scenarios does in fact detect and quantify real damage.

In general, more measurements would be expected to give better results, but the results here show that this is not always true for various reasons such as quality of signals and location of measurements with respect to the proximity to support and to excitation. Hence it is possible that the results based on two measurements can be better than those based on four measurements (as observed here). This is due to the fact that the measurements used are not the same and the quality of the signals may differ between the two sets. It may also reflect the fact that the combination of levels 2 and 6 is a good one for extracting key information from a range of modes of vibration. To observe the results of another set of two measurements, identification of the undamaged structure was carried out using measurements from levels 3 and 7. In this case, the mean false damage identified was 2.70% – larger than 1.24% and 0.96% reported above. This result reinforces the earlier discussion that levels 2 and 6 were a good combination and the likely reason is that the quality of measurements at levels 3 and 7 may not be as good as those at levels 2 and 6. Our experience has shown that it is important to have some measurement near the base of the structure. This fact also explains the poorer results obtained from the combination of levels 3 and 7 because the first two floors were not measured. In addition, the only other floor measured in this case was that at the load application point (i.e. level 7). The local vibration due to the direct application of load at level 7 could result in less meaningful information from the viewpoint of global system dynamics. In any case, the mean false damage identified remains low (<3%) in all the combinations considered.

The identification carried out for the damaged cases is summarised in tables 7.13 and 7.14. As with the case of full measurement presented previously, the results for multiple damage of different magnitude is poor. The discussion here will therefore focus on the results obtained when one or more levels are subject to the same magnitude of damage. The success achieved for these cases with the reduced measurement is displayed in figure 7.17. As expected the identification success reduces for fewer available measurements. Nevertheless, the reduction is reasonably small and the success rate achieved is still very good. Using only two measurements to identify a 7-DOF structure is encouraging, as for a more realistic structure of many DOFs it will be necessary to identify the damage with as few measurements as possible. Using only these two measurements the single large damage is consistently identified as more than double any false damage. The very small (4%) damage is also well identified with a failure rate of only 9%. In two thirds of cases the small damage is identified as more than double the maximum false damage.

7.4 Output-Only Identification

In this section, we will carry out the identification of the structure using the output-only strategy as presented in Chapter 5. Identifying damage without force measurement, for

Table 7.13 Damage detection results based on same force input for undamaged and damaged structure and incomplete measurement (1,3,5,7)

Damage Scenario		Mean damage identified (std dev)			Success (%)		
		Small	Large	Max False Damage	1X	2X	4X
Undamaged	D0			1.24 (0.42)	–	–	–
Single small	D1	3.61 (0.38)		1.67 (0.46)	89%	56%	18%
	D7	4.56 (0.67)		1.64 (0.50)	100%	80%	36%
	D8	4.07 (0.75)		1.68 (0.79)	89%	71%	29%
					93%	69%	28%
Two small	D9	4.32 (1.17)		1.73 (0.68)	89%	67%	24%
		4.56 (0.76)					
Single large	D2		17.59 (2.42)	4.34 (2.26)	100%	100%	53%
Three large	D6		18.83 (3.15)	5.10 (2.34)	100%	73%	36%
			15.47 (1.78)				
			18.89 (3.14)				
One small and one large	D3	4.64 (1.20)	17.88 (2.61)	4.11 (1.93)	64%	24%	13%
Two small and one large	D4	4.17 (1.43)	17.25 (2.38)	4.98 (2.73)	27%	20%	13%
		4.50 (1.33)					
One small and two large	D5	5.26 (2.69)	17.07 (1.41)	3.69 (3.43)	67%	31%	16%
			19.79 (2.80)				

Table 7.14 Damage detection results based on same force input for undamaged and damaged structure and incomplete measurement (2 and 6)

Damage Scenario		Mean damage identified (std dev)			Success (%)		
		Small	Large	Max False Damage	1X	2X	4X
Undamaged	D0			0.96 (0.33)	–	–	–
Single small	D1	3.13 (0.56)		1.72 (0.35)	82%	47%	4%
	D7	4.48 (0.26)		1.74 (0.82)	93%	71%	40%
	D8	4.35 (0.73)		1.20 (0.56)	98%	84%	51%
					91%	67%	32%
Two small	D9	2.94 (0.65)		1.47 (0.50)	80%	53%	16%
		4.58 (0.28)					
Single large	D2		16.04 (1.89)	5.20 (1.22)	100%	98%	33%
Three large	D6		16.53 (3.14)	5.66 (2.27)	98%	69%	22%
			18.07 (2.88)				
			18.39 (2.03)				
One small and one large	D3	2.42 (1.47)	16.75 (2.31)	5.62 (1.08)	0%	0%	0%
Two small and one large	D4	4.71 (2.59)	16.70 (2.42)	4.94 (1.44)	16%	0%	0%
		2.58 (1.82)					
One small and two large	D5	3.25 (3.79)	18.35 (3.08)	5.40 (2.71)	49%	16%	9%
			18.08 (1.60)				

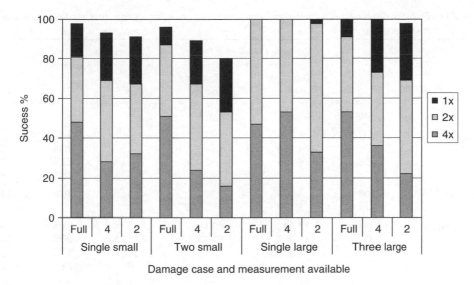

Figure 7.17 Effect of incomplete measurement on identification success

Table 7.15 Additional medium damage scenarios

New damage scenario	Undamaged case	Damaged case	Resulting medium damage
D10	D1	D2	Level 4
D11	D4	D5	Level 6
D12	D5	D6	Level 3
D13	D4	D6	Levels 3 and 6

this case where structural modelling error is inevitable, poses a significant challenge. In order to assess the effectiveness of the strategy, in addition to the 9 damage scenarios considered in the previous section, four additional damage scenarios of 'medium' 13% damage are considered as indicated in table 7.15. Following the findings of the identification using force measurements, only cases where the same input forces are used to identify both the damaged and undamaged structures are considered here.

Identification is carried out using the same 500 data points obtained for the previous tests, except now only the acceleration measurements are used while force measurements are ignored. The mass is treated as known and is set as 3.4 kg for each storey. It is noted that the mass is not exact but is treated as a reasonable estimate based on the results of the previous tests. The GA parameters used are the same as those used for the damaged (known mass) structure, as given in table 7.10. Identification is first carried out using full acceleration measurements and the computational time, for identification of each structure, is 114 s on a Pentium 4, 3-GHz PC. An overview of the results is given in table 7.16. While the results are reasonably good for the large damage cases, the strategy is unable to consistently detect the small damage cases, and

Table 7.16 Damage detection results for no force measurement using a single test and full acceleration measurement

Damage Case		Damage Identified		Success %		
		Damaged level	Max false damage	1X	2X	4X
Undamaged	D0	–	6.077	–	–	–
Single small	D1	4.034	4.418	51%	33%	2%
	D7	4.950	5.193	58%	20%	0%
	D8	4.754	6.551	51%	22%	9%
				53%	25%	4%
Two small	D9	4.896	5.975	51%	20%	4%
		6.199				
Single medium	D10	13.648	6.718	82%	58%	36%
	D11	15.068	5.283	91%	80%	51%
	D12	14.828	3.113	100%	87%	69%
				91%	75%	52%
Two medium	D13	15.582	4.772	84%	71%	40%
		15.551				
Single large	D2	17.127	7.102	91%	69%	44%
Three large	D6	20.909	4.609	98%	87%	53%
		17.375				
		19.578				

Each damage case is tested 45 times.

even medium damage is occasionally missed. The reason for this comes from the fact that the strategy uses a model based updating of the force. As the structural model does not exactly match the physical model, errors in structural identification are induced. The effect of these errors can be reduced, however, by combining the results of several tests. This procedure is demonstrated in the following subsection.

The force identified is also of interest. Figure 7.18 shows a comparison of the identified force and the force measured by the load cell during the experiment. The agreement between the identified forceand and measured force is good, thereby suggesting that this method is good in identifying not only structural parameters but also external forces.

7.4.1 Using Multiple Test Data

In order to reduce the uncertainty in the identification results, it is proposed that data from several tests be combined to more accurately identify structural parameters. In order to do this, the calculation of force and simulation of accelerations is carried out separately for each set of test data, using the trial structural parameters. The total sum-square-error over all tests is then used to determine the fitness of the solution. In this way we do not simply find the parameters that best fit a single test, but find parameters that give the best overall result for a number of experimental tests.

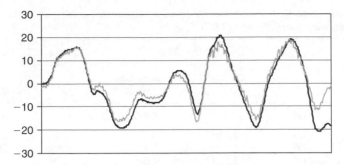

Figure 7.18 Comparison of identified force (Heavy line = measured force, light line = identified force)

Table 7.17 Damage detection results for no force measurement using two tests and full acceleration measurement

Damage Case		Damage Identified		Success %		
		Damaged level	Max false damage	1X	2X	4X
Undamaged	D0	–	4.167	–	–	–
Single small	D1	3.914	2.787	74%	40%	7%
	D7	4.561	2.991	86%	32%	11%
	D8	4.244	3.946	62%	34%	6%
				74%	53%	8%
Two small	D9	4.647 5.185	2.912	78%	42%	12%
Single medium	D10	13.948	4.847	98%	83%	37%
	D11	15.623	3.335	100%	98%	68%
	D12	14.210	2.230	100%	99%	90%
				99%	93%	65%
Two medium	D13	15.719 16.076	2.649	100%	96%	76%
Single large	D2	17.327 19.959	4.961 3.560	99% 100%	87% 98%	50% 64%
Three large	D6	17.575 19.625				

Each damage case is tested 90 times.

This procedure is first applied using two tests and full acceleration measurements. Here the two tests used have different force profiles, for example, forces A and B, or forces B and C etc. The same two force profiles are used for the identification of the structure before and after damage has taken place. All possible force combinations are considered, resulting in a total of 90 test cases. The important results are given in table 7.17.

Table 7.18 Damage detection results for no force measurement using two tests and incomplete acceleration measurement (2,6,7)

| Damage Case | | Damage Identified | | Success % | | |
		Damaged level	Max false damage	1X	2X	4X
Undamaged	D0	–	4.942	–	–	–
Single small	D1	3.626	3.674	53%	21%	4%
	D7	4.147	3.925	59%	20%	1%
	D8	3.554	4.045	53%	12%	3%
				55%	18%	3%
Two small	D9	3.363	3.530	54%	24%	10%
		5.299				
Single medium	D10	14.388	5.923	96%	70%	27%
	D11	15.588	3.718	100%	98%	60%
	D12	13.812	2.549	100%	99%	86%
				99%	89%	58%
Two medium	D13	15.733	2.915	100%	97%	73%
		16.594				
Single large	D2	17.514	6.630	98%	74%	31%
Three large	D6	20.545	4.107	99%	94%	57%
		17.388				
		19.831				

Each damage case is tested 90 times.

While the computational time needed to run the additional data is doubled, the results are far more reliable. All of the medium and large damage cases are now identified with almost 100% success. Even when only three acceleration measurements are available the identification success remains good. These results are summarised in table 7.18. This idea of combining multiple test data in order to average out uncertainties could be applied to any case including the tests with measured force presented earlier in this chapter. This is achieved at a cost of increased computational time, but for situations where accuracy is of most importance, this option is essential.

To gain some insight into the accuracy that can be achieved if more data is available, identification is carried out using 5, and then 15 sets of data, where measurement is only available at three locations. The results achieved for 5 tests, shown in table 7.19, are averaged over nine cases that were considered for each damage scenario. The result for 15 data sets (all of the tests conducted) is shown in table 7.20. Here, as all available data is used, the test could only be completed once and success can no longer be reported as a percentage of successful trials. Instead, the success reported in table 7.20 is the ratio of the smallest real damage identified to the maximum false damage.

When five data sets are used, even the single small damage can be identified in most cases, and for 15 data sets, even the very difficult cases of multiple damage magnitude are correctly identified. While it may not always be practical to run so many tests, it is encouraging to know that, given sufficient information, we are able to accurately

Table 7.19 Damage detection results for no force measurement using five tests and incomplete acceleration measurement (2,6,7)

Damage Case		Damage Identified		Success (out of 9)		
		Damaged	Max false	1X	2X	4X
Undamaged	D0	–	2.217	–	–	–
Single small	D1	3.611	1.987	9/9	4/9	0/9
	D7	4.044	2.197	9/9	5/9	0/9
	D8	3.014	2.532	6/9	3/9	0/9
				88%	44%	0%
Two small	D9	3.594	2.050	6/9	3/9	0/9
		4.903		67%	33%	0%
Single medium	D10	13.937	4.615	9/9	8/9	2/9
	D11	15.637	2.673	9/9	9/9	9/9
	D12	14.078	1.834	9/9	9/9	9/9
				100%	96%	74%
Two medium	D13	16.374	1.688	9/9	9/9	9/9
		16.824		100%	100%	100%
Single large	D2	17.047	4.334	9/9	9/9	5/9
				100%	100%	56%
Three large	D6	20.830	2.631	9/9	9/9	9/9
		17.035		100%	100%	100%
		20.094				

Each damage case is tested only 9 times. Success is therefore given as $n/9$, where n is the number of runs which achieved the given level of success.

identify damage for these 'output only' cases, even with only a limited number of sensors.

A comparison of the results achieved is shown in figure 7.19. The maximum false damage identified for the case of a single small damage is used for the comparison. The false damage is closely related to identification success and gives us a good measure of the reliability of the results. The figure shows how important the multiple test option is in achieving reasonable levels of false damage. For a single test the maximum false damage is of the same magnitude as the damage to be identified and we cannot separate the damage. Nevertheless, even with limited measurements, the maximum false damage is reduced to 2.2% and 1.1% for 5 and 15 tests, respectively, and we gain more confidence in the identified damage. For a real case we would need to balance the computational time available with the required accuracy. If constrained by very limited time and budget, a single test would quickly identify large damage, but would not be able to confidently detect small changes in stiffness. With more time (both experimental and numerical) and a few extra tests, the confidence in detecting damage of even very small level can be increased considerably.

Table 7.20 Damage detection results for no force measurement using 15 tests and incomplete acceleration measurement (2,6,7)

Damage Case		Identified damage							Success
		1	2	3	4	5	6	7	
Single 4%	D1	−0.25	0.68	0.99	3.62	−0.81	−0.18	0.23	3.66
	D7	−2.98	1.14	0.47	1.28	−0.38	4.00	−0.52	3.13
	D8	0.26	0.65	2.96	−0.47	−0.38	0.93	0.63	3.18
Two 4%	D9	−2.72	1.78	3.41	0.82	−0.76	4.89	0.11	1.92
Single 13%	D10	4.61	−0.46	0.83	13.95	−0.38	−0.79	0.32	3.03
	D11	−1.00	−1.46	2.85	0.75	−0.45	15.68	0.03	5.50
	D12	−1.71	0.40	13.99	−1.58	−0.09	1.37	1.54	9.08
Two 13%	D13	−2.73	−1.05	16.44	−0.82	−0.54	16.84	1.57	10.47
Single 17%	D2	4.37	0.22	1.82	17.06	−1.19	−0.98	0.54	3.90
Three 17%	D6	−0.91	0.97	20.76	17.07	−2.52	20.14	2.21	7.72
Multiple	D3	1.51	1.36	2.27	18.12	−1.58	3.06	0.02	1.35
Damage	D4	1.77	2.01	5.17	17.74	−1.96	3.96	0.65	1.97
	D5	0.79	0.58	7.87	18.36	−2.42	19.03	0.68	9.96

Damaged levels highlighted in bold and shaded. Max false damage is in bold. Success is reported as the smallest real damage/max false damage.

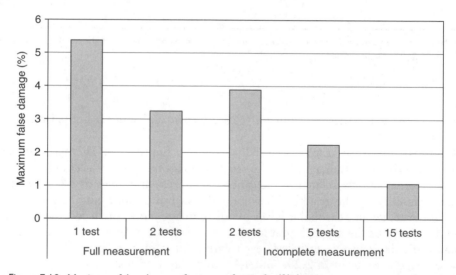

Figure 7.19 Maximum false damage for case of a single 4% damage

7.5 Chapter Summary

Experimental studies are essential in providing a realistic test of numerical strategies. Unfortunately they have often been lacking from similar research due to the much greater difficulty they pose than numerical simulation studies. To this end, the damage detection strategy based on the SSRM developed in the previous chapter has been verified in the damage detection of a seven-storey steel model. The strategy makes use of measurement of both the undamaged and damaged structures to significantly improve the accuracy and reliability of detection. The strategy uses measurement of the undamaged structure in order to calibrate the structural model, and by fixing the mass of the structure based on the identification of the undamaged structure, is able to reduce the damage detection step to a simpler known mass problem. The parameters identified during the calibration step are also used to initiate the search when identifying the damaged structure.

The damage level considered in the experiments (only 4%) is significantly smaller than the 10–50% damage generally assumed, even in numerical studies. These tests have helped to assess the performance of the strategy and also to identify some practical issues that exist. The model calibration in particular has proved important to reduce modelling error and results have shown that the same input force should be used when identifying the undamaged structure (calibration) and the damaged structure (damage detection). This is essential, particularly when small levels of damage are to be identified. The detection results presented are excellent when a single magnitude of damage is to be detected. Single or multiple damaged levels with 4% damage are identified in almost 100% of cases, an impressive result considering the experimental noise level was estimated as 1–10%. In detecting a combination of large and small damage, the modelling error could cause the identification of a false damage to be of similar magnitude to the small real damage. In these cases the large damage may be easily identified but the small damage is not. Identification using a reduced number of acceleration measurements has also been presented. The experimental results have shown the strategy to be very robust in this respect. With only four, or even two available measurements, the success rate is very high. Using only two measurements, a single damage of 4% is identified in 91% of cases. In two thirds of cases the identified 4% damage is more than double any false damage identified.

The output-only time domain strategy presented earlier has also been validated. The strategy works by simultaneously computing the excitation forces as the structural parameters (stiffness and damping) are identified. The experimental test results once again highlight the importance of a good structural model. As the strategy uses the dynamic equation of motion to update the forces, the results are dependant on how well the structure and the numerical model match. The results show that this problem can be significantly reduced by utilising data from several tests. For example, if five tests are used, the maximum false damage is reduced to merely 2.2% thereby enabling identification of even small levels of damage. Larger damage of 10–20% is easily identified using one or two tests and is identified with great certainty if more tests are available. This idea of multiple data sets could also be applied in a general case, for example, when force measurements are available to further improve the reliability of identification.

Chapter 8

Substructure Methods of Identification

The numerical difficulty in obtaining accurate results increases dramatically with the number of unknown parameters in the system of interest. In this chapter, the strategy of "divide-and-conquer" is presented to address this issue, by means of the substructure concept. The key idea is to divide the structure into substructures such that the number of unknown parameters is within manageable size in each stage of identification. Called the substructural identification (SSI), this method enables us to identify part of the structure if other parts are not required for identification. Even if the whole structure is to be identified, using SSI to identify different parts of the structure in stages will significantly (or even drastically) improve the computational accuracy and speed due to the smaller system size in terms of the numbers of DOFs and unknowns involved at the substructure level.

The first paper on substructural system identification was by Koh et al. (1991) using the extended Kalman filter method to identify unknown structural parameters. Further work was presented by Su and Juang (1994) on the procedures for substructure state-space models, assembling substructure transfer function data and deduction of substructure Markov parameters. Yun and Lee (1997) proposed a substructural identification method using the sequential prediction error method and an auto-regressive and moving average with stochastic input model. Subsequent research works adopting the substructural approach include those by Oreta and Tanabe (1994), Hermann and Pradlwarter (1998), Yun and Bahng (2000), Koh et al. (2000, 2003a), Koh and Shankar (2003a,b), Tee et al. (2005) and Huang and Yang (2008).

Though this divide-and-conquer idea seems straightforward, special attention is needed in the formulation because substructure is not isolated from the remainder of the structure (or adjacent substructures). It is essential to account for interaction forces at interface between the substructure and other parts of the structure. The improved GA strategy as presented in the earlier chapters is employed to facilitate the divide-and-conquer identification strategy. A numerical simulation study is presented, including a fairly large system of 100 DOFs to illustrate the identification performance. The methods are tested for known-mass as well as unknown-mass systems with 202 unknown parameters. Effects of incomplete and noisy measurements are accounted for.

8.1 Substructural Identification

In order to illustrate the substructural identification strategy, consider a lumped mass system as shown in figure 8.1(a). The substructure to be identified is shown in

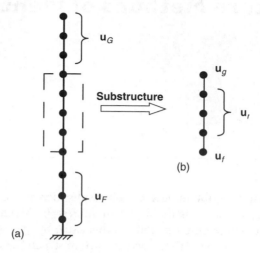

Figure 8.1 (a) Complete structure (b) A substructure

figure 8.1(b). To derive the equations of motion for the substructure, the equations of motion for the complete structure can be written in the following partitioned form.

$$
\begin{bmatrix}
\mathbf{M}_{FF} & \mathbf{M}_{Ff} & & & \\
\mathbf{M}_{fF} & \mathbf{M}_{ff} & \mathbf{M}_{fr} & & \\
& \mathbf{M}_{rf} & \mathbf{M}_{rr} & \mathbf{M}_{rg} & \\
& & \mathbf{M}_{gr} & \mathbf{M}_{gg} & \mathbf{M}_{gG} \\
& & & \mathbf{M}_{Gg} & \mathbf{M}_{GG}
\end{bmatrix}
\begin{Bmatrix}
\ddot{u}_F \\ \ddot{u}_f \\ \ddot{u}_r \\ \ddot{u}_g \\ \ddot{u}_G
\end{Bmatrix}
+
\begin{bmatrix}
\mathbf{C}_{FF} & \mathbf{C}_{Ff} & & & \\
\mathbf{C}_{fF} & \mathbf{C}_{ff} & \mathbf{C}_{fr} & & \\
& \mathbf{C}_{rf} & \mathbf{C}_{rr} & \mathbf{C}_{rg} & \\
& & \mathbf{C}_{gr} & \mathbf{C}_{gg} & \mathbf{C}_{gG} \\
& & & \mathbf{C}_{Gg} & \mathbf{C}_{GG}
\end{bmatrix}
\begin{Bmatrix}
\dot{u}_F \\ \dot{u}_f \\ \dot{u}_r \\ \dot{u}_g \\ \dot{u}_G
\end{Bmatrix}
$$

$$
+
\begin{bmatrix}
\mathbf{K}_{FF} & \mathbf{K}_{Ff} & & & \\
\mathbf{K}_{fF} & \mathbf{K}_{ff} & \mathbf{K}_{fr} & & \\
& \mathbf{K}_{rf} & \mathbf{K}_{rr} & \mathbf{K}_{rg} & \\
& & \mathbf{K}_{gr} & \mathbf{K}_{gg} & \mathbf{K}_{gG} \\
& & & \mathbf{K}_{Gg} & \mathbf{K}_{GG}
\end{bmatrix}
\begin{Bmatrix}
u_F \\ u_f \\ u_r \\ u_g \\ u_G
\end{Bmatrix}
=
\begin{Bmatrix}
P_F \\ P_f \\ P_r \\ P_g \\ P_G
\end{Bmatrix}
\tag{8.1}
$$

In the above equation, subscript 'r' denotes internal DOFs of the substructure concerned, subscripts 'F' and 'G' denote the DOFs of the remaining structure on the two sides marked as F and G in figure 8.1(a). Subscripts 'f' and 'g' denote interface DOFs of the substructure with the remaining structure on the two sides F and G, respectively. For conciseness, let subscript j denote all interface DOFs (i.e. $j = f \cup g$). For the substructure considered, the equations of motion may be extracted from the above equation system to yield

$$
[\mathbf{M}_{rj} \quad \mathbf{M}_{rr}]
\begin{Bmatrix}
\ddot{u}_j(t) \\ \ddot{u}_r(t)
\end{Bmatrix}
+ [\mathbf{C}_{rj} \quad \mathbf{C}_{rr}]
\begin{Bmatrix}
\dot{u}_j(t) \\ \dot{u}_r(t)
\end{Bmatrix}
+ [\mathbf{K}_{rj} \quad \mathbf{K}_{rr}]
\begin{Bmatrix}
u_j(t) \\ u_r(t)
\end{Bmatrix}
= \mathbf{P}_r(t)
\tag{8.2}
$$

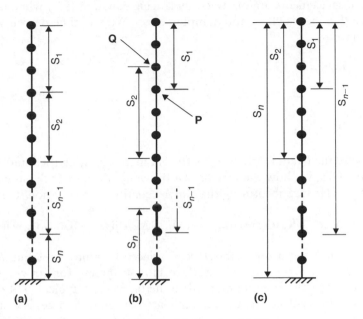

Figure 8.2   (a) SSI without Overlap; (b) SSI with Overlap; (c) PSI

Treating interaction effects at the interface ends as "input", the above equation system can be organised as

$$\mathbf{M}_{rr}\ddot{\mathbf{u}}_r(t) + \mathbf{C}_{rr}\dot{\mathbf{u}}_r(t) + \mathbf{K}_{rr}\mathbf{u}_r(t) = \mathbf{P}_r(t) \quad \mathbf{M}_{rj}\ddot{\mathbf{u}}_j(t) - \mathbf{C}_{rj}\dot{\mathbf{u}}_j(t) - \mathbf{K}_{rj}\mathbf{u}_j(t) \qquad (8.3)$$

Two versions of the SSI approach are possible depending on whether there is any overlap between adjacent substructures.

8.1.1   *Substructural Identification Without Overlap*

The first version of substructural identification is the *SSI without overlap* version as illustrated in figure 8.2(a). Adjacent substructures do not overlap, i.e. no structural member appears in more than one substructure. The identification procedure requires response measurements at interface DOFs which are treated as input to the substructure of concern. In principle, equation (8.3) can be used for SSI (Koh and See 1991) but this will require accelerations, velocities and displacements at the interface DOFs as evident in the RHS of the equation. For practical reason, acceleration signals measured by means of accelerometers is normally preferred over velocity and displacement signals. To eliminate the requirement of displacement and velocity signals which contain errors if numerically integrated from acceleration signals, the concept of "quasi-static displacement" vector is adopted. The displacements for internal DOFs are split into quasi-static displacements ($\mathbf{u}_r^s$) and "relative" displacements ($\mathbf{u}_r^*$), i.e.

$$\mathbf{u}_r(t) = \mathbf{u}_r^s(t) + \mathbf{u}_r^*(t) \qquad (8.4)$$

Quasi-static displacements are obtained by solving equation (8.3) while ignoring the applied force ($\mathbf{P}_r$), inertia effect and damping effect. With all time-derivative terms set to zero, we have

$$\mathbf{K}_{rr}\mathbf{u}_r^s = -\mathbf{K}_{rj}\mathbf{u}_j \tag{8.5}$$

Hence,

$$\mathbf{u}_r^s = -\mathbf{K}_{rr}^{-1}\mathbf{K}_{rj}\mathbf{u}_j = \mathbf{r}\,\mathbf{u}_j \tag{8.6}$$

where $\mathbf{r}$ is called the (static) influence coefficient matrix relating the internal DOFs to the interface DOFs, i.e. how the internal DOFs would displace if the interface DOFs are statically displaced. Substituting the above equation into equation (8.3) leads to

$$\mathbf{M}_{rr}\ddot{\mathbf{u}}_r^*(t) + \mathbf{C}_{rr}\dot{\mathbf{u}}_r^*(t) + \mathbf{K}_{rr}\mathbf{u}_r^*(t) = \mathbf{P}_r(t) - (\mathbf{M}_{rj} + \mathbf{M}_{rr}\mathbf{r})\ddot{\mathbf{u}}_j(t) - (\mathbf{C}_{rj} + \mathbf{C}_{rr}\mathbf{r})\,\dot{\mathbf{u}}_j(t) \tag{8.7}$$

The RHS without $\mathbf{P}_r$ term represents forces induced by motion relating to interface DOFs and may be referred to as "interface motion forces" for convenience. Since damping force is usually small compared to inertia force in typical civil engineering structures, the velocity-dependent part in the interface motion forces is assumed to be negligible. Hence,

$$\mathbf{M}_{rr}\ddot{\mathbf{u}}_r^*(t) + \mathbf{C}_{rr}\dot{\mathbf{u}}_r^*(t) + \mathbf{K}_{rr}\mathbf{u}_r^*(t) = \mathbf{P}_r(t) - (\mathbf{M}_{rj} + \mathbf{M}_{rr}\mathbf{r})\ddot{\mathbf{u}}_j(t) \tag{8.8}$$

Consequently, only accelerations (instead of displacements or velocities) at interface DOFs are required to compute the interface motion forces. Furthermore, if there is no excitation within the substructure, $\mathbf{P}_r$ simply vanishes and the method can advantageously be used for "output-only" identification (i.e. no force measurement is necessary) for identification of the substructure.

The forward analysis as required in the GA approach involves solving the above equations of motion subjected to the excitation and interface motion forces. Note that the fitness function is defined in terms of the measured components of the relative accelerations ($\ddot{\mathbf{u}}_r^*$). A block diagram of the GA-implementation for the SSI method is presented in figure 8.3.

For lumped mass systems, $\mathbf{M}_{rj}$ vanishes and equation (8.8) can be simplified to

$$\mathbf{M}_{rr}\ddot{\mathbf{u}}_r^*(t) + \mathbf{C}_{rr}\dot{\mathbf{u}}_r^*(t) + \mathbf{K}_{rr}\mathbf{u}_r^*(t) = \mathbf{P}_r(t) - \mathbf{M}_{rr}\mathbf{r}\ddot{\mathbf{u}}_j(t) \tag{8.9}$$

In addition, if a substructure includes the free end, such as substructure S_1 in figure 2(a), the influence coefficients matrix reduces to simply

$$\mathbf{r} = [1 \quad 1 \quad \cdots \quad 1]^T \tag{8.10}$$

8.1.2 Substructural Identification With Overlap

Another version of substructural identification is the *SSI with overlap* version as illustrated in figure 8.2(b). Adjacent substructures are allowed to have overlap members,

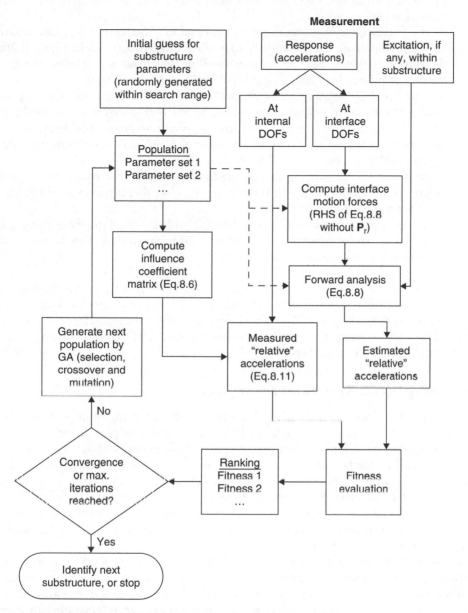

Figure 8.3 Block diagram for GA-based SSI method

i.e. some structural members appear in two substructures (or more). The overlap has the following four implications.

(1)   A node (or DOF) may be interface to one substructure but internal to another substructure. For example, for identification of substructure S_1 in figure 8.2(b), acceleration measurement at interface node A is required as input. Nevertheless,

for S_2-identification, node A is an internal node and its measured response is treated as output.

(2)   At the overlap, interface acceleration required in substructure S_{i+1} can actually be computed from the previously identified substructure S_i. In figure 8.2(b), the upper interface node of S_2 is node B at which no acceleration measurement is required. Instead, having identified S_1, the acceleration at node A can be computed by solving the equations of motion for S_1. Thus, less response measurements are required as compared to SSI without overlap. Nevertheless, error propagation may result since any identification error of S_1 is carried forward to S_2 in SSI with overlap, whereas the identification of one substructure is independent from others in SSI without overlap.

(3)   Once identified in a substructure S_i, the overlap member can optionally be taken as known in the subsequent substructure S_{i+1} but this gives another source of error propagation.

(4)   While interface masses are excluded in SSI without overlap, these masses are included in one substructure or another in SSI with overlap. This facilitates the identification of all masses if unknown.

Despite the above differences, the GA-implementation procedure for the SSI with overlap is very similar to that for the SSI without overlap. In fact, the block diagram as shown in figure 8.3 still applies. The main deviation is that some interface accelerations are not necessarily obtained from measurement but numerically computed in a previous substructure as explained earlier.

8.1.3 *Progressive Structural Identification*

In general, identification results would improve with increasing number of response measurements (given that the measurement locations of course play an influential role too). In the above-mentioned SSI approach, only measurements within the substructure of concern and at interface ends are used when identifying the substructure parameters. Given a limited number of sensors, the identification results can be enhanced greatly if the measurement program allows shift of measurement sensors according to the substructure under investigation. If this is not possible, an alternative is to make use of as many response measurements as possible by progressively expanding the domain of substructural identification.

To this end, a variation of the SSI called the progressive structural identification (PSI) method is presented as illustrated in figure 8.2(c). This employs the same idea of dividing the system to be identified into several sub-systems to improve the identification performance. The main difference is that the substructure grows progressively while still keeping the number of unknowns small at each stage. For the structure as shown in figure 2(c), the topmost substructure is selected as the first substructure (S_1) for identification. After its unknown structural parameters are identified, an extended substructure is considered, i.e. the second substructure (S_2) which includes the previous substructure (S_1). But the identified parameters of S_1 are taken as known. The response signals used in substructure S_1 are used again in the fitness evaluation while identifying the parameters of S_2. This procedure continues until the whole or required part of the structure is identified.

As compared to the SSI method, the main advantage of the PSI method is that it utilizes increasing availability of response measurements without increasing the number of unknowns as the substructure enlarges. The cost incurred is the increase in computational time for the forward analysis involving solving larger systems. This shortcoming may, nevertheless, be offset by faster convergence at each stage due to the availability of more response measurements.

8.2 Numerical Examples

8.2.1 Identification of 100-DOF Known-Mass System

To check the applicability and performance of the proposed methods, numerical simulation study is carried out on examples with known exact values. A 100-DOF known mass structural system is used to observe the performance of the SSI method without overlap. For comparison purpose, the GA method is also applied to the *complete* structure in a direct manner and is referred here as the "complete structural identification" (CSI) method.

The 100-DOF lumped mass system similar to that presented in figure 8.1 is considered. The aim here is to identify the 100 stiffness parameters. The exact parameters are: $m_1 = 6000\,\text{kg}$, $m_2 = m_3 \ldots = m_{100} = 3000\,\text{kg}$, $k_1 = k_2 \ldots k_{100} = 7000\,\text{kN/m}$. Rayleigh damping is assumed and modal damping ratio of 2% is used for modes 1 and 10. A lower damping is used compared to the earlier example to ensure the damping in the higher modes is not too high, as for the substructure method the higher frequency modes are important and should not be numerically damped out. Though not the main focus, the damping constants are included as unknown in all the identification examples. The numerically simulated response is obtained for 1s at time step of 0.002 s (500 steps) by using Newmark's constant-acceleration method and random noise is added. The improved GA strategy is applied for both the SSI and CSI methods with the parameters shown in table 8.1. The reduced data length procedure is used with 200 (40%) of the data used for 50% of the generations. The computational time is approximately 12 minutes for the SSI method and 25 minutes for the CSI. It is important to note that although 10 substructures are analysed and the total evaluations is much greater, the SSI takes a much shorter time than the CSI due to the much smaller systems to be simulated.

In order to further demonstrate the performance of the SSI a third set of tests is conducted with a larger population and number of generations. The identification

Table 8.1 Identification parameters used – Known mass

	CSI	SSI	SSI (with larger population)
Number of substructures	1	10	10
Population size	20 × 3	10 × 3	20 × 3
Runs	4/10	4/10	4/10
Generations	200	150	200
Crossover rate	0.8	0.8	0.8
Mutation rate	0.2	0.2	0.2
Regeneration	3	3	3
Reintroduction	50	50	50

parameters here are the same as for the CSI. This case has a computational time of 30 minutes, which is only slightly longer than the CSI despite the fact that 10 substructures are analysed using the same number of evaluations as the CSI structure. Ten excitation forces act on the 10th, 20th ... and 100th nodes. Acceleration response measurements (simulated) are assumed to be available at 40 nodes, viz. the 2nd, 5th, 7th and 10th nodes in each set of 10 nodes. The same force and response measurements are used so that a fair comparison of the methods can be made.

(1)　*CSI*: The complete structure is identified as a whole without substructuring.
(2)　*SSI without overlap*: The structure is divided into ten substructures. The first substructure begins from node 90 to node 100, but the mass at node 90 (where acceleration is used as input) is not included in the substructure formulation. This substructure is denoted as $S_1 = [90–100]$ for convenience. Square bracket indicates that acceleration measurement at interface node 90 is required (as input). Parenthesis indicates that acceleration measurement at node 100 is optional and, if available (as in this case), is treated as output. The second substructure is $S_2 = [80–90]$, for which the measurement at interface nodes 80 and 90 are treated as input. The other substructures are treated in the same way right down to the base node 0. Note that S_1 is used for identification of k_{91} to k_{100}, whereas S_2 is for k_{81} to k_{90} and so on.

Due to the stochastic nature of the GA approach, the comparison is performed using 10 sets of identification results corresponding to different initial populations which are randomly generated within the specified search range of 0.5 to 1.5 times the actual values. It is noted that the *relative* accelerations are used in the fitness evaluation and are derived from Eqs. (8.4) and (8.6) as follows:

$$\ddot{\mathbf{u}}_r^*(t) = \ddot{\mathbf{u}}_r(t) - \mathbf{r}\ddot{\mathbf{u}}_j(t) \tag{8.11}$$

Table 8.2 presents the mean and maximum absolute errors of identified stiffness parameters corresponding to each data set of initial population for all the methods considered. The results consistently show that the divide-and-conquer idea indeed works. The results appear to improve less at larger noise levels, however this is likely due to the fact that the best possible results will be limited at this higher noise. Also the

Table 8.2 Average Identification Results – Known mass

	CSI	SSI	SSI (with larger population)
0% Noise			
Mean Absolute Error	1.23	0.32	0.10
Maximum Absolute Error	5.70	2.46	0.92
5% Noise			
Mean Absolute Error	1.53	0.83	0.68
Maximum Absolute Error	5.98	3.65	2.66
10% Noise			
Mean Absolute Error	1.74	1.59	1.37
Maximum Absolute Error	6.79	6.80	6.11

large number of measurements available to the CSI will help more at larger noise levels whereas the SSI only has 4 measurements to use for identification in each substructure. Notwithstanding the above the results are excellent considering the size of the system and a relatively small identification time. It is also a credit to the main identification procedure that even the CSI is also able to give reasonable results on this very large system.

8.2.2 Identification of 100-DOF Unknown-Mass System

As further demonstration of the substructure method, we consider the same 100-DOF structure as the previous example but with all mass, stiffness and damping parameters assumed unknown. In this example the focus is to identify a specific part of the structure (floors 60–70) without having to determine information from the rest of the structure. For this purpose excitation is provided only at floors 60, 65 and 70 and accelerations measured at floors 60, 61, 63, 65, 67, 69 and 70. As with the previous example, acceleration data is available for 500 data points at a sampling rate of 0.002s. This example illustrates perhaps the greatest strength of the substructure method; the ability to identify part of a structure without any requirement on the remainder of the structure. A substructure with internal nodes 61-69 is considered, allowing for identification of stiffness k_{61} to k_{70} and mass from m_{61} to m_{69}. The accelerations measured at nodes 60 and 70 are used as input at the boundaries of substructure. The results from an average of 10 runs using the GA parameters of table 8.3 are presented in table 8.4. Approximate computational time for the identification is 29 minutes. The

Table 8.3 Identification parameters used – Unknown Mass (internal nodes 61–69 substructure)

Population size	60×3
Runs	5/15
Generations	360
Crossover rate	0.4
Mutation rate	0.2
Regeneration	3
Reintroduction	120

Table 8.4 Average Identification Results – Unknown Mass (internal nodes 61–69 substructure)

	Stiffness	Mass
0% Noise		
Mean Absolute Error	0.57	0.54
Maximum Absolute Error	1.02	1.04
5% Noise		
Mean Absolute Error	0.89	0.84
Maximum Absolute Error	1.79	1.62
10% Noise		
Mean Absolute Error	3.21	3.10
Maximum Absolute Error	5.59	5.79

results illustrate the exceptional ability of the substructure method, in combination with the SSRM, to quickly and accurately identify required information from even very large unknown mass systems. The ability of the substructure method to isolate a part of the structure is one of the most promising solutions available for tackling large real life problems with many hundreds of nodes.

8.3 Chapter Summary

A practical divide-and-conquer strategy by means of substructure concept is presented for identification of large systems with many unknowns. The numerical difficulty in identifying large systems is overcome by working in smaller and more manageable search domains. The improved GA strategy is readily adopted as the identification tool for its robustness and efficiency. Note that, to some extent, the results presented have accounted for modelling error associated with damping. This is because the damping term in the interface motion force is neglected as explained earlier. Furthermore, the damping constants of the Rayleigh model adopted for the full structure are not necessarily the same as those at the substructure level. In spite of such modelling errors, the identified results are generally very good. The salient features of each of the methods considered are compared below and summarized in Table 8.5.

(1) *CSI*: By applying GA-based identification to the complete structure, this method is relatively easy to implement but the identification accuracy or efficiency may not be acceptable for large systems.

(2) *SSI without overlap* – The identification of each substructure is completely independent of other substructures. The identification sequence of substructures is immaterial, and it is thus also possible to identify only part of the structure

Table 8.5 Pros and Cons of Different Identification Methods

Method	Pros	Cons
Complete structural identification	• Easy to implement – no substructural formulation is needed	• Search domain involves all unknown parameters, making convergence difficult and time consuming for large systems
Substructural identification without overlap	• No interface motion force from adjacent substructures is required • Easy to implement parallel computing at substructure level	• If needed, interface masses have to be identified after all other parameters are identified
Substructural identification with overlap	• Applicable to identifying unknown mass • Less response measurements if making using of acceleration solution obtained in another substructure	• Interface motion forces from adjacent substructures are required • Error propagation if result from one substructure is used in another substructure
Progressive structural identification	• No interface motion force from adjacent substructures is required • Applicable to identifying unknown mass	• Error propagation as result from one substructure is used in the next substructure

without the knowledge of others. As a result, there is no error propagation problem and parallel computing can be easily applied. Furthermore, if the excitation force is outside the substructure, no force measurement is required leading to an "output-only" identification approach. Nevertheless, interface masses are not included in identification of any substructure. If interface masses are to be identified, it is necessary to refine substructures so as to include the interface masses yet to be determined, or conduct CSI with only these interface masses as unknown.

(3)   *SSI with overlap* – Some measurements are used both as interface measurements (input) to one substructure and as internal measurements (output) to another. Thus the number of measurements required can be reduced as compared to SSI without overlap. Besides, identification of all masses including interface masses is possible since they are included in one substructure or another. Nevertheless, error propagation may arise from inaccuracy if (a) some interface measurements are computed from previously identified substructures, or (b) structural parameters of overlap members if assumed as known in the subsequent substructure. In such cases, the identification of each substructure is to some extent dependent on the identification sequence and the results of previous substructures.

(4)   *PSI* – The main idea is to include as many available measurements as possible while keeping the number of unknowns small at each stage of identification. Interface masses are included in stages. As a result, the PSI method may be the most suitable for identification of unknown-mass systems. The computational cost for forward analysis increases as the substructure grows in size. Nevertheless, this may be offset by a faster convergence due to the use of more response measurements and the total computational time is thus not necessarily much longer than SSI.

References

ABAQUS (1998), Standard User's Manual (Version 5.8), Hibbit, Karlsson and Sorensen, Inc., Pawtucket, Rhode Island, USA.

Adams, R.D., Cawley, P., Pye, C.J. and Stone, B.J. (1978), "A vibration technique for non-destructively assessing the integrity of structures." Journal of Mechanical Engineering Science, 20 (2), 93–100.

Adeli, H. (2001), "Special Issue: Health Monitoring of Structures." Computer-Aided Civil and Infrastructure Engineering, 16(1).

Adeli, J. and Karim, A. (2000), "Fuzzy-wavelet RBFNN model for freeway incident detection." Journal of Transportation Engineering, ASCE, 126(6), 464–471.

Aktan, A.E., Lee, K.H., Chuntavan, C. and Aksel, T. (1994), Modal testing for structural identification and condition assessment of constructed facilities, in Proceedings of the 12th International Modal Analysis Conference, Honolulu, Hawaii, 462–468.

Axelrod, R. (1987), "The evolution of strategies in the itterated prisoner's dilemma," in Davis, L. (Ed), Genetic Algorithms and Simulated Annealing, Pitman, London, 32–41.

Bagley, J.D. (1967), The behavior of adaptive systems which employ genetic and correlation algorithms, Doctoral Dissertation, University of Michigan.

Baker, J.E. (1987), Reducing Bias and Inefficiency in the Selection Algorithm, in Grefenstette, J.J. (Ed), Genetic Algorithms and their Applications: Proceedings of the Second International Conference on Genetic Algorithms, 14–21.

Beck, J. L. and Katafygiotis, L. S. (1998). "Updating models and their uncertainties: Bayesian statistical framework." Journal of Engineering Mechanics, 124(4), 455–461.

Begambre, O. and Laier, J. E. (2009). "A hybrid Particle Swarm Optimization – Simplex algorithm (PSOS) for structural damage identification." Advances in Engineering Software, 40(9), 883–891.

Bernal, D. and Beck, J. (2004), "Special Issue: Phase 1 of the IASC-ASCE Structural Health Monitoring Benchmark." Journal of Engineering Mechanics, ASCE, 130(1).

Bicanic, N. and Chen, H. P. (1997). "Damage identification in framed structures using natural frequencies." International Journal for Numerical Methods in Engineering, 40(23), 4451–4468.

Booker, L. (1987), Improving Search in Genetic Algorithms, in Davis, L. (Ed), Genetic Algorithms and Simulated Annealing, Pitman, London, 61–73.

Caravani, P., Watson, M.L. and Thomson, W.T. (1977), "Recursive Least-Squares Time Domain Identification of Structural Parameters." Journal of Applied Mechanics, ASME, 44, 135–140.

Carden, E.P. and Fanning, P. (2004), "Vibration Based Condition Monitoring: A Review, Structural Health Monitoring." 3, 355–377.

Catbas, F.N., Susoy, M., and Frangopol, D.M. (2008). "Structural health monitoring and reliability estimation: Long span truss bridge application with environmental monitoring data." Engineering Structures, Elsevier, 30(9), 2347–2359.

Carmichael, D.G. (1979), "The State Estimation Problem in Experimental Structural Mechanics." Proceedings of 3rd International Conference on Applications of Stastics and Probability in Soil and Structural Engineering, Sydney, 802–815.

Chan, T.H.T., Yu, L., Tam, H.Y., Ni, Y.Q., Liu, S.Y., Chung, W.H., and Cheng, L.K. (2006), "Fiber Bragg grating sensors for structural health monitoring of Tsing Ma Bridge: background and experimental observation." Engineering Structures, Vol. 28, No. 5, 648–659.

Chang, C.C. (2007). "Special issue: management of civil infrastructure." Structure and Infrastructure Engineering Engineering, Vol. 3, No. 2, 93–185.

Chang, P.C., Flatau, A. and Liu, S.C. (2003), "Review Paper: Health Monitoring of Civil Infrastructure." Structural Health Monitoring 2, 257–267.

Chen, H. P. and Bicanic, N. (2000). "Assessment of damage in continuum structures based on incomplete modal information." Computers and Structures, 74(5), 559–570.

Chen, S., Billings, S.A. and Grant, P.M. (1990), "Non-linear system identification using neural networks." International Journal of Control 51(6), 1191–1214.

Chou, J.H. and Ghaboussi, J. (2001), "Genetic algorithm in structural damage detection, Computers and Structures." 79, 1335–1353.

Cooley, J.W. and Tukey, J.W. (1965), "An Algorithm for the Machine Calculation of Complex Fourier Series." Mathematics of Computation 19(90), 297–311.

Creed, S.G. (1987), "Assessment of large engineering structures using data collected during in-service loading." in Garas, F.K. Clarke, J.L. and Armer, G.S.T. (Ed), Structural Assessment: the use of full and large scale testing, Butterworths, London, 55–62.

De Jong, K.A. (1975), An analysis of the behavior of a class of genetic adaptive systems, Doctoral Dissertation, University of Michigan.

DiPasquale, E., and Cakmak, A. S. (1988). Identification of the serviceability limit state and direction of seismic structural damage, Tech. Rep. NCEER-88-0022, Nat. Ctr. for Earthquake Engrg. Res., Buffalo. N.Y.

Farrar, C.R. and Doebling, S.W. (1997), "Lessons Learned from Applications of Vibration-Based Damage Identification Methods to a Large Bridge Structure, in Structural Health Monitoring – Current Status and Perspectives." Proceedings of the International Workshop on Structural Health Monitoring, Stanford University, Sept 18–20, 351–370.

Farrar, C.R. and Worden, K. (2007). "Theme Issue on Structural Health Monitoring." Philosophical Transactions of the Royal Society A, 365 (1851).

Fogel, D.B. (1998), Evolutionary Computation: The Fossil Record, IEEE Press, New York.

Franco, G., Betti, R. and Lus, H. (2004), "Identification of Structural Systems Using an Evolutionary Strategy." Journal of Engineering Mechanics, ASCE, 130(10), 1125–1139.

Frangopol, D.M. and Messervey, T.B. (2009). "Life-cycle cost and performance prediction: Role of structural health monitoring." Chapter 16 in Frontier Technologies for Infrastructures Engineering, S-S, Chen and A.H-S. Ang, eds., Structures and Infrastructures Book Series, Vol. 4, D. M. Frangopol, Book Series Editor, CRC Press/Balkema, Boca Raton, London, New York, Leiden, 361-381.

Frangopol, D.M., and Messervey, T.B. (2009). "Maintenance principles for civil structures," Chapter 89 in Encyclopedia of Structural Health Monitoring, C. Boller, F-K. Chang, and Y. Fujino, eds., John Wiley & Sons Ltd, Chicester, UK, Vol. 4, 1533–1562.

Friswell, M.I. (2007). "Damage Identification Using Inverse Methods." Philosophical Transactions of the Royal Society A, 365 (1851), 393–410.

Fryba, L. and Pirner, M. (2001), "Load tests and modal analysis of bridges." Engineering Structures 23, 102–109.

Furuta, H., Kameda, T., Nakahara, K., Takahashi, Y. and Frangopol, D. M. (2006), "Optimal bridge maintenance planning using improved multi-objective genetic algorithm." Structure and Infrastructure Engineering, Taylor & Francis, 2(1), 33–41.

Ghanem, R. and Shinozuka, M. (1995), "Structural System Identification. I: Theory." Journal of Engineering Mechanics, ASCE, 121(2), 255–263.

Ghanem, R. and Sture, S. (2000). "Special Issue: Structural Health Monitoring." Journal of Engineering Mechanics, ASCE, 126(7).

Goldberg, D.E. and Lingle, R. (1985), Alleles, Loci, and the Travelling Salesman Problem, in Grefenstette, J. (Ed), Proceedings of an International Conference on Genetic Algorithms and their Applications, Lawrence Erlbaum Associates, 154–159.

Goldberg, D.E. (1989), Genetic Algorithms in Search, Optimization, and Machine Learning, Addison-Wesley Publishing Company.

Grefenstette, J., Gopal, R., Rosmaita, B. and Van Gucht, D. (1985), Genetic Algorithms for the Travelling Salesman Problem, in Grefenstette, J. (Ed), Proceedings of an International Conference on Genetic Algorithms and their Applications, Lawrence Erlbaum Associates, 160–168.

He, R. S. and Hwang, S. F. (2006). "Damage detection by an adaptive real-parameter simulated annealing genetic algorithm." Computers & Structures, 84(31–32), 2231–2243.

Herrmann, T. and Pradlwarter, H.J. (1998). "Two-step Identification Approach for Damped Finite Element Models." Journal of Engineering Mechanics, ASCE, 124 (6), 639–647.

Hjelmstad, K. D. and Shin, S. (1997). "Damage Detection and Assessment of Structures from Static Response." Journal of Engineering Mechanics, ASCE, 123(6), 568–576.

Holland, J.H. (1962a), "Information Processing in Adaptive Systems, in Information Processing in the Nervous System." Proceedings of the International Union of Physiological Sciences, XXII International Congress, Leiden.

Holland, J.H. (1962b), "Outline for a Logical Theory of Adaptive Systems." Journal of the Association for Computing Machinery, 9, 297–314.

Holland, J.H. (1968), Hierarchical Descriptions, Universal Spaces and Adaptive Systems, Technical Report, OAR Projects 01252 and 08226, University of Michigan.

Holland, J.H. (1971), "Processing and Processors for Schemata." in Jacks, E.L. (Ed), Associative Information Processing, 127–146, New York: American Elsevier.

Holland, J.H. (1973), "Genetic Algorithms and the Optimal Allocation of Trials." SIAM Journal on Computing 2(2), 88–105.

Holland, J.H. (1975), Adaptation in Natural and Artificial Systems, University of Michigan Press, Ann Arbour.

Hoshiya, M. and Saito, E. (1984), "Structural Identification by Extended Kalman Filter." Journal of Engineering Mechanics, ASCE, 110(12), 1757–1770.

Hsieh, K.H., Halling, M.W. and Barr, P.J. (2006), "Overview of Vibrational Structural Health Monitoring with Representitive Case Studies." Journal of Bridge Engineering, ASCE, 11(6), 707–715.

Humar, J., Bagchi, A. and Xu, H. (2006), "Performance of Vibration-Based Techniques for the Identification of Structural Damage." Structural Health Monitoring, 5, 215–242.

Huang, H.W. and Yang, J.N. (2008), "Damage identificaion of substructure for local health monitoring." Smart Structures and Systems, 4(6), 795–807.

Iba, H., Kurita, T., De Garis, H. and Sato,T. (1993), "System Identification using Structured Genetic Algorithms." in Forrest, S. (Ed), Proceedings of the fifth International Conference on Genetic Algorithms, Morgan Kaufmann, California, 279–286.

Imai, H., Yun, C. B. and Shinozuka, M. (1989). "Fundamentals of system identification in structural dynamics." Probabilistic Engineering Mechanics, 4(4), 162–173.

Jiang, S.-F., C.-M. Zhang and Koh, C.G. (2006) "Structural damage detection by integrating data fusion and probabilistic neural network." Advances in Structural Engineering, 9(4), 445–458.

Kalman, R.E. (1960), "A New Approach to Linear Filtering and Prediction Problems." Journal of Basic Engineering, ASME, 82(Series D), 35–45.

Kim, J.T. and Stubbs, N. (1995), "Model-Uncertainty Impact and Damage-Detection Accuracy in Plate Girder." Journal of Structural Engineering, ASCE, 121 (10), 1409–1417.

Kishore Kumar, R., Sandesh S. and Shankar, K. (2007). "Parametric identification of non-linear dynamic systems using Levenburg-Marquardt method and genetic algorithm." International Journal of Structural Stability and Dynamics, 7(4), 715–725.

Kitagawa, G. (1996). "Monte Carlo filter and smoother for non-Gaussian state space models." Journal of Computational and Graphical Statistics, 5(1), 1–25.

Ko, J.M., and Ni, Y.Q. (2005), "Technology developments in structural health monitoring of large-scale bridges." Engineering Structures, 27(12), 1715–1725.

Ko, J.M., Ni, Y.Q. and Chan, T.H.T. (1999). "Dynamic monitoring of structural health in cable-supported bridges." Proceedings of International Symposium on Smart Systems for Bridges, Structures and Highways, S.C. Liu ed., New Port Beach, USA, March, 370–381.

Koehler, G.J. (1997), New directions in genetic algorithm theory, Annals of operations research, 75, pg 49–68.

Koh, C.G., See, M. and Balendra, T. (1991), "Estimation of structural parameters in time domain: a substructure approach." Earthquake Engineering and Structural Dynamics, 20(8), 787–801.

Koh, C.G., Hong, B and Liaw, C.Y. (2000), "Parameter Identification of Large Structural Systems in Time Domain." Journal of Structural Engineering, 126(8), 957–963.

Koh, C.G., Hong, B and Liaw, C.Y. (2003a), "Substructural and progerssive structural identification methods." Engineering Structures, 25, 1551–1563.

Koh, C.G., Chen, Y.F. and Liaw, C.Y. (2003b), "A Hybrid Computational Strategy for Identification of Structural Parameters." Computers and Structures, 81, 107–117.

Koh, C.G. and See, M. (1994), "Identification and Uncertainty Estimation of Structural Parameters, Journal of Engineering Mechanics." 120 (6), 1219–1236.

Koh, C.G. and See, M. (1999), "Techniques in the identification and uncertainty estimation of parameter in structural systems." in Leondes, C.T (Ed), Structural Dynamic Systems Computational Techniques and Optimixation, Computational Techniques, Gordon and Breach Science Publishers.

Koh, C.G. and Shankar, K. (2003a), "Stiffness Identification by a Substructural Approach in Frequency Domain." International Journal of Structural Stability and Dynamics 3(2), 267–281.

Koh, C.G. and Shankar, K. (2003b), "Substructural Identification Method Without Interface Measurement." Journal of Engineering Mechanics, ASCE, 129(7), 769–776.

Koh, H-M., and Frangopol, D.M., eds. (2008). Bridge Maintenance, Safety, Management, Health Monitoring and Informatics, Set of Book and CD-ROM, A Balkema Book (ISBN 13: 978-0-415-46844-2 (hbk), 786 pages) and CD-ROM (ISBN 13 978-0-415-46844-2), 465 full length papers, CRC Press, Taylor & Francis Group, Boca Raton, London, New York, Leiden, 2008.

Lee, J. (2009). "Identification of multiple cracks in a beam using natural frequencies." Journal of Sound and Vibration, 320(3), 482–490.

Li, L., Yang, Y., Peng, H. and Wang X (2006), "Parameters identification of chaotic systems via chaotic ant swarm." Chaos, Solitons and Fractals, 28(5), 1204–1211.

Ling, X. and Haldar, A. (2004), "Element Level System Identification with Unknown Input with Rayleigh Damping." Journal of Engineering Mechanics, ASCE, 30 (8), 877–885.

Liu, G.R. and Chen, S.C. (2002). "A novel technique for inverse identification of distributed stiffness factor in structures." Journal of Sound and Vibration, 254(5), 823–835.

Liu, M., Frangopol, D.M. and Kim, S. (2009a). "Bridge system performance assessment from structural health monitoring: A case study." Journal of Structural Engineering, ASCE 135(6), 733–742.

Liu, M., Frangopol, D.M. and Kim, S. (2009b). "Bridge safety evaluation based on monitored live load effects." Journal of Bridge Engineering, ASCE, 14(4), 257–269.

Ljung, L. (1986). "Frequency and time domain methods in system identification." Modeling identification and robust control, C.I. Byrnes and A. Lindquist, eds., 615–624.

Ljung, L. and Glover, K. (1981), "Frequency Domain Versus Time Domain Methods in System Identification." Automatica, 17 (1), 71–86

Luh, G.C. and Wu, C.Y. (1999), "Non-linear system identification using genetic algorithms." Proceedings of the Institution of Mechanical Engineers, 213(1), 105–117.

Mangal, L., Idichandy, V.G. and Ganapathy, C. (2001), "Structural monitoring of offshore platforms using impulse and relaxation response." Ocean Engineering, 28, 689–705.

Maybeck, P.S. (1979), Stochastic Models, Estimation, and Control, Volume 1, Academic Press, New York.

Michalewicz, Z. (1994), Genetic Algorithms + Data Structures = Evolution Programs, Second, Extended Edition, Springer-Verlag, Berlin Heidelberg.

Ni, Y.Q., Ko, J.M., and Zheng, G. (2002), "Dynamic analysis of large-diameter sagged cables taking into account flexural rigidity." Journal of Sound and Vibration, Vol. 257, No. 2, 301–319.

Okasha, M.N. and Frangopol, D.M. (2009). "Lifetime-oriented multi-objective optimization of structural maintenance, considering system reliability, redundancy, and life-cycle cost using GA." Structural Safety, Elsevier, 31(6), 460–474.

Oreta, W. C. and Tanabe, T. A. (1994), "Element Identification of Member Properties of Framed Structures." Journal of Structural Engineering, ASCE, 120(7), 1961–1976.

Perera, R. and Torres, R. (2005), "Structural Damage Assessment using Genetic Algorithms." 9th International Confernce on Inspection, Appraisal, Repairs & Maintenance of Structures, 20–21.

Perry, M.J., Koh, C.G. and Choo, Y.S. (2006), "Modified Genetic Algorithm Strategy for Structural Identification." Computers and Structures, 84, 529–540.

Potts, J.C., Giddens, T.J. and Yadav, S.B. (1994), "The Development and Evaluation of an Improved Genetic Algorithm Based on Migration and Artificial Selection." IEEE Trans. on Systems Man. and Cybernetics, 24 (1), 73–86.

Raghavendrachar, M. and Aktan, A.E. (1992), "Flexability by Multireference Impact Testing for Bridge Diagnosis." Journal of Structural Engineering, ASCE, 118 (8), 2186–2203.

Rao, M.A., Srinivas, J. and Murthy, B.S.N. (2004), "Damage detection in vibrating bodies using genetic algorithms." Computers and Structures, 82, 963–968.

Roberts, J.B. and Vasta, M. (2000), "Parametric Identification of Systems with Non-Gaussian Excitation using Measured Response Spectra." Probabilistic Engineering Mechanics, 15, 59–71.

Sanayei, M. and Onipede, O. (1991), "Damage assessment of structures using static test data." AIAA Journal, 29(7), 1174–1179.

Salawu, O.S. (1997), "Detection of Structural Damage through Changes in Frequency: a Review." Engineering Structures 19 (9), 718–723.

Salawu, O.S. and Williams, C. (1995), "Bridge Assessment Using Forced Vibration Testing." Journal of Structural Engineering 121(2), 161–173.

Sato, T., and Kaji, K. (2000). "Structural System Identification using Monte Carlo Filter." 3rd US-Japan workshop on nonlinear system identification and structural health monitoring.

Sawyer, J.P. and Rao S.S. (2000), "Structural damage detection and identification using fuzzy logic." AIAA Journal 38(12), 2328–2335.

Schaffer, J.D., Caruana, R.A., Eshelman, L.J and Das, R. (1989), "A Study of Control Parameters Affecting Online Performance of genetic Algorithms for Function Optimization." Third International Conference on Genetic Algorithms, George Mason University, June 4–7.

Shi, T., Jones, N.P. and Ellis, J.H. (2000), "Simultaneous Estimation of System and Input Parameters from Output Measurements." Journal of Engineering Mechanics, ASCE, 126(7), 746–753.

Shinozuka, M. and Ghanem, R. (1995), "Structural System Identification. II: Experimental Verification." Journal of Engineering Mechanics, ASCE, 121 (2), 265–273.

Shinozuka, M., Yun, C. B. and Imai, H. (1982). "Identification of linear structural dynamic systems." Proceedings of the American Society of Civil Engineer, ASCE, 108(EM6), 1371–1389.

Spanos, P.D. and Lu, R. (1995), "Nonlinear System Identification in Offshore Structural Reliability." Journal of Offshore Engineering and Artic Engineering, ASME, 117, 171–177.

Su, T.-J. and Juang, J.-N. (1994). "Substructure System Identification and Synthesis." J. Guidance, Control, and Dynamics, 17(5), 1087–1095.

Tang, H. Fukuda, M. and Xue, S. (2007). "Particle swarm optimization for structural system identification." The 6th International Workshop on Structural Health Monitoring, Stanford, CA, 483–492.

Tang, H. S. Xue, S. T. and Fan, C. X. (2008). "Differential evolution strategy for structural system identification." Computers & Structures, 86(21–22), 2004–2012.

Tee, K.F., Koh, C.G. and Quek, S.T. (2005), "Substructural First and Second Order Model Identification for Structural Damage Assessment." Earthquake Engineering and Structural Dynamics, 34(15), 1755–1775.

Topping, B.H.V. and Tsompanakis, Y. (2009), Proceedings of the First International Conference on Soft Computing Technology in Civil, Structural and Environmental Engineering, Civil-Comp Press, Stirling, U.K.

Tsai, C.H. and Hsu, D.S. (1999), "Damage Diagnostics of Existing Reinforced Concrete Structures." in Kumar, B. and Topping, B.H.V (Ed), Artificial Intelegence Applications in Civil and Structural Engineering, Civil-Comp Press, Edinburgh, 85–92.

Vanik, M.W., Beck, J.L., and Au, S.K. (2000). "Bayesian probabilistic approach to structural health monitoring." Journal of Engineering Mechanics, 126(7), 738–745.

Wahab, M.M.A. and De Roeck, G. (1999), "Damage Detection in Bridges Using Modal Curvatures: Application to a Real Damage Scenario." Journal of Sound and Vibration, 226 (2), 217–235.

Wang, X., Hu, N., Fukunaga, H. and Yao Z.H. (2001), "Structural damage identification using static test data and changes in frequencies." Engineering Structures, 23(6), 610–621.

Welch, G. and Bishop, G. (2004), An Introduction to the Kalman Filter, UNC-Chapel Hill, TR 95-041, April 5, 2004.

Whitley, D. (1989), "The GENITOR Algorithm and Selective Pressure: Why Rank-Based Allocation of Reproductive Trials is Best." Proceedings of the Third International Conference on Genetic Algorithms, George Mason University, June 4–7, 116–121.

Wu, Z. and Fujino, Y. (2006). "Special Issue: Structural Health Monitoring and Intelligient Infrastrcutre." Smart Materials and Structures, 14(3).

Yeung, W.T. and Smith, J.W. (2005). "Damage Detection in Bridges using Neural Networks for Pattern Recognition of Vibration Signatures." Engineering Structures, 27, 685–698.

Yoshida, I. and Sato, T. (2002). "Health Monitoring Algorithm by the Monte Carlo Filter Based on Non-Gaussian Noise." Journal of Natural Disaster Science, 20 (2), 101–107.

Yuen, K.V. and Katafygiotis, L.S. (2001). "Bayesian time-domain approach for modal updating using ambient data." Probabilistic Engineering Mechanics, 16(3), 219–231.

Yuen, K.V., Au, S.K. and Beck, J.L. (2004). "Two-stage structural health monitoring approach for phase I benchmark studies." Journal of Engineering Mechanics, 130(1), 16–33.

Yuen, K.V. and Katafygiotis, L. S. (2006). "Substructure identification and health monitoring using noisy response measurements only." Computer-Aided Civil and Infrastructure Engineering, 21(4), 280–291.

Yun, C.B. and Bahng, E.Y. (2000), "Substructural Identification Using Neural Networks." Computer and Structures, 77(1), 41–52.

Yun, C.B. and Lee, H.-J. (1997). "Substructural Identification for Damage Estimation of Structures." J. Structural Safety, 19(1), 121–140.

Yun, C. B., and Shinozuka, M., (1980), "Identification of nonlinear structural dynamic system." Journal of Structural Mechanics, ASCE, 8(2), 187–203.

Zhao, J. and DeWolf, T. (1999), "Sensitivity Study for Vibrational Parameters Used in Damage Detection." Journal of Structural Engineering, ASCE, 125 (4), 410–416.

Appendix

A.I   A Simple GA Code

The following shows a simple GA program that finds the maximum value of a given function as discussed in chapter 3. The function value is assumed positive over the search range specified and is defined in the sub function $f(x)$. The program here is coded in FORTRAN.

Input file

The input is named in.txt in the following form.

```
N         LL       UL
Pop_size  Tot_gen
P_cross   P_mut
```

PROGRAM Simple_GA

```fortran
IMPLICIT NONE                                              ! Variable Declaration
INTEGER:: i,j,stat                                        ! counters, file open status
INTEGER:: g,Tot_gen                                       ! generation number and total generations
INTEGER:: N, Pop_size                                     ! number of bits, population size
INTEGER:: cross                                           ! crossover location
REAL(8):: x, P_cross, P_mut                               ! x, Crossover and mutation probabilities
REAL(8):: LL, UL, r                                       ! search limits, random number
REAL(8), DIMENSION(:), ALLOCATABLE:: Fitness              ! fitness of individuals
INTEGER, DIMENSION(:), ALLOCATABLE:: Select               ! vector for crossover operation
INTEGER, DIMENSION(:), ALLOCATABLE:: Twos                 ! vector to store the powers of two
REAL(8), DIMENSION(:), ALLOCATABLE:: P_select             ! vector to store selection probabilities
INTEGER, DIMENSION(:,:), ALLOCATABLE:: Pop, T_Pop         ! population, temporary pop
INTEGER, DIMENSION(:), ALLOCATABLE:: O1, O2, best         ! offspring for crossover, best solution
REAL(8):: Best_fit=0

OPEN(UNIT=1,FILE='in.txt',STATUS="OLD",IOSTAT=stat)       ! open input file, in.txt
IF (stat/=0) STOP "*** ERROR OPENING INPUT FILE ***"      ! check file opened successfully
READ(1,*) N, LL, UL                                       ! read inputs from file
READ(1,*) Pop_size, Tot_gen
READ(1,*) P_cross, P_mut
CLOSE (1)                                                 ! close input file

ALLOCATE(Pop(Pop_size,N), T_pop(Pop_size,N))              ! allocate matrices and vectors
ALLOCATE(Fitness(Pop_size))
ALLOCATE(P_select(Pop_size), Select(Pop_size))
ALLOCATE(O1(N), O2(N), Twos(N), Best(N))
```

```fortran
Twos(N)=1                              ! compute powers of 2
DO i=1,N-1                             ! for binary to real conversion
    Twos(N-i)=Twos(N+1-i)*2
END DO


                                       ! Generate initial population
CALL RANDOM_SEED()                     ! assigns the seed for random numbers
Pop=0                                  ! set all values to 0 initially
DO i=1, Pop_size
    DO j=1, N
        CALL RANDOM_NUMBER(r)          ! generates r in range [0 1]
        IF (r>0.5) Pop(i,j)=1          ! decide if bit should be 0 or 1
    END DO
END DO


DO g=1, Tot_gen                        ! Main analysis loop

  DO i=1, Pop_size                     ! Evaluate fitness
      x=SUM(pop(i,1:N)*twos)           ! converts the binary number to an integer
      x=LL+(UL-LL)*x/(2**N-1)          ! converts integer to real x value
      Fitness(i)=f(x)                  ! f(x) defined in function subprogram below
      IF (Fitness(i)>Best_fit) THEN
          Best_fit=fitness(i)          ! Store best result
          Best=Pop(i,1:N)
      END IF
  END DO
  IF (g==Tot_gen) EXIT                 ! final gen - skip crossover and mutation

  P_select=Fitness/SUM(Fitness)        ! Selection - Roulette wheel selection
  DO i=2,Pop_size                      ! proportional to fitness
    P_select(i)=P_select(i)+P_select(i-1)   ! cumulative probability
  END DO


  DO i=1,Pop_size
    CALL RANDOM_NUMBER(r)              ! r decides which solutions are reproduced
    DO j=1,Pop_size
      IF (P_select(j)>=r) THEN
          T_Pop(i,1:N)=Pop(j,1:N)      ! new population stored temporarily
          EXIT
      END IF
    END DO
  END DO
  Pop=T_Pop


  j=0                                  ! Crossover
  DO i=1,Pop_size                      ! j to record the number of individuals selected
    CALL RANDOM_NUMBER(r)
    IF (r<P_cross) THEN
```

```
      j=j+1
      Select(j)=i                              ! record which individual selected
    END IF
  END DO
  CALL Shuffle(Select(1:j),j)                  ! Shuffles the selected individuals
  IF (MOD(j,2)==1) j=j-1                        ! if odd number selected remove one
  DO i=1,j,2
    CALL RANDOM_NUMBER(r)                       ! randomly select crossover location
    cross=CEILING(r*(N-1))
    O1(1:cross)=Pop(Select(i),1:cross)
    O2(1:cross)=Pop(Select(i+1),1:cross)
    O1(cross+1:N)=Pop(Select(i+1),cross+1:N)
    O2(cross+1:N)=Pop(Select(i),cross+1:N)
    Pop(Select(i),1:N)=O1
    Pop(Select(i+1),1:N)=O2
  END DO

  DO i=1, Pop_size                             ! Mutation
    DO j=1, N
      CALL RANDOM_NUMBER(r)
      IF(r<P_mut) THEN
        IF (Pop(i,j)==0) THEN
          Pop(i,j)=1
        ELSE
          Pop(i,j)=0
        END IF
      END IF
    END DO
  END DO

END DO                                          ! End main analysis loop
x=SUM(Best*twos)                                ! converts the best result to an integer
x=LL+(UL-LL)*x/(2**N-1)                          ! converts integer to real x value
PRINT*, "max value of", Best_fit, "Was found at x= ", x   ! Output final result to screen

STOP

CONTAINS
FUNCTION F(x)                                   ! Function for computing f(x)

REAL(8) :: F
REAL(8), INTENT(IN):: x

F=0.5-((sin(2*x))**2-0.5)/(1.0+0.02*x*x)         ! Example function used in Chapter 2

END FUNCTION f                                  ! End of function f(x)

SUBROUTINE Shuffle(vector, size)                ! Subroutine to shuffle vector
```

```
    IMPLICIT NONE
    INTEGER,INTENT(IN) :: size
    INTEGER,INTENT(INOUT) :: vector(size)
    INTEGER :: temp(size),i,loc
    REAL :: shuf(size)

    CALL RANDOM_NUMBER(shuf)
    DO i=1,size
       loc=MAXLOC(shuf,DIM=1)
       temp(i)=vector(loc)
       shuf(loc)=0
    END DO
    vector=temp
```

```
END SUBROUTINE Shuffle                     ! End of shuffle subroutine

END PROGRAM Simple_GA                       ! End of program
```

A.2   SDOF Identification

The following is a sample program for SDOF identification as discussed in Chapter 4.

Input file

The input file is named in.txt in the following form.

```
N          LL        UL
Pop_size   Tot_gen
P_cross    P_mut
L h m c
```

PROGRAM Simple_GA_SDOF

```
    IMPLICIT NONE                                          ! Variable Declaration
    INTEGER:: i,j,stat                                     ! counters, file open status
    INTEGER:: g,Tot_gen                                    ! generation number and total generations
    INTEGER:: N, Pop_size                                  ! number of bits, population size
    INTEGER:: cross                                        ! crossover location
    REAL(8):: P_cross, P_mut                               ! Crossover and mutation probabilities
    REAL(8):: LL, UL, r                                    ! search limits, random number
    REAL(8), DIMENSION(:), ALLOCATABLE:: Fitness           ! fitness of individuals
    INTEGER, DIMENSION(:), ALLOCATABLE:: Select            ! vector for crossover operation
    INTEGER, DIMENSION(:), ALLOCATABLE:: Twos              ! vector to store the powers of two
    REAL(8), DIMENSION(:), ALLOCATABLE:: P_select          ! vector to store selection probabilities
    INTEGER, DIMENSION(:,:), ALLOCATABLE:: Pop, T_Pop      ! population, temporary pop
    INTEGER, DIMENSION(:), ALLOCATABLE:: O1, O2, best      ! offspring for crossover, best solution
    REAL(8):: Best_fit=0

    REAL(8):: x, v, a, delx                                ! response
    REAL(8):: k, m, c                                      ! structural parameters
```

```fortran
REAL(8):: h, sse                                      ! time step, error
INTEGER:: L,t                                         ! length of response data, step number
REAL(8), DIMENSION(:), ALLOCATABLE:: a_s, a_m, F      ! simulated and measured accelerations,
                                                      ! forces


OPEN(UNIT=1,FILE='in.txt',STATUS="OLD",IOSTAT=stat)   ! open input file, in.txt
IF (stat/=0) STOP "*** ERROR OPENING INPUT FILE ***"  ! check file opened successfully
READ(1,*) N, LL, UL                                   ! read inputs from file
READ(1,*) Pop_size, Tot_gen
READ(1,*) P_cross, P_mut
READ(1,*) L, h, m, c
ALLOCATE(a_s(L),a_m(L),F(L))
DO i=1,L
READ(1,*) F(I)
END DO
DO i=1,L
READ(1,*) a_m(i)
END DO
CLOSE (1)                                             ! close input file


ALLOCATE(Pop(Pop_size,N), T_pop(Pop_size,N))          ! allocate matrices and vectors
ALLOCATE(Fitness(Pop_size))
ALLOCATE(P_select(Pop_size), Select(Pop_size))
ALLOCATE(O1(N), O2(N), Twos(N), Best(N))


Twos(N)=1                                             ! compute powers of 2
DO i=1,N-1                                            ! for binary to real conversion
Twos(N-i)=Twos(N+1-i)*2
END DO

                                                      ! Generate initial population
CALL RANDOM_SEED()                                    ! assigns the seed for random numbers
Pop=0                                                 ! set all values to 0 initially
DO i=1, Pop_size
DO j=1, N
    CALL RANDOM_NUMBER(r)                             ! generates r in range [0 1]
    IF (r>0.5) Pop(i,j)=1                             ! decide if bit should be 0 or 1
END DO
END DO


DO g=1, Tot_gen                                       ! Main analysis loop


DO i=1, Pop_size                                      ! Evaluate fitness
    k=SUM(pop(i,1:N)*twos)                            ! converts the binary number to an integer
    k=LL+(UL-LL)*k/(2**N-1)                           ! converts integer to real x value

    x=0
    v=0
    a=0
```

```
DO t=1,L
  delx=(F(t)+m*a+(c+4*m/h)*v-k*x)/(4*m/(h*h)+2*c/h+k)
  x=x+delx
  v=2*delx/h-v
  a=(F(t)-c*v-k*x)/m
  a_s(t)=a
END DO
sse=SUM((a_s-a_m)**2)
Fitness(i)=1.0/(0.001+sse)

IF (Fitness(i)>Best_fit) THEN
    Best_fit=fitness(i)                    ! Store best result
    Best=Pop(i,1:N)
  END IF
END DO
IF (g==Tot_gen) EXIT                        ! final gen - skip crossover and mutation

P_select=Fitness/SUM(Fitness)              ! Selection - Roulette wheel selection
DO i=2,Pop_size                            ! proportional to fitness
    P_select(i)=P_select(i)+P_select(i-1)  ! cumulative probability
END DO

DO i=1,Pop_size
  CALL RANDOM_NUMBER(r)                     ! r decides which solutions are reproduced
  DO j=1,Pop_size
    IF (P_select(j)>=r) THEN
      T_Pop(i,1:N)=Pop(j,1:N)              ! new population stored temporarily
      EXIT
    END IF
  END DO
END DO
Pop=T_Pop

j=0                                        ! Crossover
DO i=1,Pop_size                            ! j to record the number of individuals selected
    CALL RANDOM_NUMBER(r)
    IF (r<P_cross) THEN
      j=j+1
      Select(j)=i                          ! record which individual selected
    END IF
END DO
CALL Shuffle(Select(1:j),j)                ! Shuffles the selected individuals
IF (MOD(j,2)==1) j=j-1                      ! if odd number selected remove one
DO i=1,j,2
    CALL RANDOM_NUMBER(r)                   ! randomly select crossover location
    cross=CEILING(r*(N-1))
    O1(1:cross)=Pop(Select(i),1:cross)
    O2(1:cross)=Pop(Select(i+1),1:cross)
    O1(cross+1:N)=Pop(Select(i+1),cross+1:N)
```

```
        O2(cross+1:N)=Pop(Select(i),cross+1:N)
        Pop(Select(i),1:N)=O1
        Pop(Select(i+1),1:N)=O2
    END DO

    DO i=1, Pop_size                          ! Mutation
        DO j=1, N
            CALL RANDOM_NUMBER(r)
            IF(r<P_mut) THEN
                IF (Pop(i,j)==0) THEN
                    Pop(i,j)=1
                ELSE
                    Pop(i,j)=0
                END IF
            END IF
        END DO
    END DO

    END DO                                    ! End main analysis loop

    k=SUM(Best*twos)                          ! converts the best result to an integer
    k=LL+(UL-LL)*k/(2**N-1)                    ! converts integer to real x value

PRINT*, "max fitness of", Best_fit, "Was found at k= ", k   ! Output final result to screen

STOP

CONTAINS

    SUBROUTINE Shuffle(vector, size)          ! Subroutine to shuffle vector

    IMPLICIT NONE
    INTEGER,INTENT(IN) :: size
    INTEGER,INTENT(INOUT) :: vector(size)
    INTEGER :: temp(size),i,loc
    REAL :: shuf(size)

    CALL RANDOM_NUMBER(shuf)
    DO i=1,size
        loc=MAXLOC(shuf,DIM=1)
        temp(i)=vector(loc)
        shuf(loc)=0
    END DO
    vector=temp

    END SUBROUTINE Shuffle

END PROGRAM Simple_GA_SDOF                     ! End of program
```

A.3   Newmark's Constant Average Acceleration Method

Newmark's method works directly on the general dynamic equilibrium equation

$$\mathbf{M\ddot{x} + C\dot{x} + Kx = F} \tag{A.1}$$

Acceleration is assumed to be constant over each time step h, from time step k to $k+1$

$$\ddot{\mathbf{x}} = \frac{\ddot{\mathbf{x}}_k + \ddot{\mathbf{x}}_{k+1}}{2} \tag{A.2}$$

Integrating twice with respect to time,

$$\dot{\mathbf{x}}_{k+1} = \dot{\mathbf{x}}_k + \left(\frac{\ddot{\mathbf{x}}_k + \ddot{\mathbf{x}}_{k+1}}{2}\right)h \tag{A.3}$$

$$\mathbf{x}_{k+1} = \mathbf{x}_k + \dot{\mathbf{x}}_k h + \left(\frac{\ddot{\mathbf{x}}_k + \ddot{\mathbf{x}}_{k+1}}{2}\right)\frac{h^2}{2} \tag{A.4}$$

Rearranging the above equations leads to representations for incremental acceleration and incremental velocity as

$$\Delta\ddot{\mathbf{x}} = \ddot{\mathbf{x}}_{k+1} - \ddot{\mathbf{x}}_k = \frac{4\Delta\mathbf{x}}{h^2} - \frac{4\dot{\mathbf{x}}_k}{h} - 2\ddot{\mathbf{x}}_k \tag{A.5}$$

$$\Delta\dot{\mathbf{x}} = \dot{\mathbf{x}}_{k+1} - \dot{\mathbf{x}}_k = \frac{2\Delta\mathbf{x}}{h} - 2\dot{\mathbf{x}}_k \tag{A.6}$$

Substitution into the equilibrium equation at time step $k+1$ gives,

$$\mathbf{M}\{\ddot{\mathbf{x}}_k + \Delta\ddot{\mathbf{x}}\} + \mathbf{C}\{\dot{\mathbf{x}}_k + \Delta\dot{\mathbf{x}}\} + \mathbf{K}\{\mathbf{x}_k + \Delta\mathbf{x}\} = \mathbf{F}_{k+1} \tag{A.7}$$

$$\mathbf{M}\left\{\ddot{\mathbf{x}}_k + \left(\frac{4\Delta\mathbf{x}}{h^2} - \frac{4\dot{\mathbf{x}}_k}{h} - 2\ddot{\mathbf{x}}_k\right)\right\} + \mathbf{C}\left\{\dot{\mathbf{x}}_k + \left(\frac{2\Delta\mathbf{x}}{h} - 2\dot{\mathbf{x}}_k\right)\right\} + \mathbf{K}\{\mathbf{x}_k + \Delta\mathbf{x}\} = \mathbf{F}_{k+1} \tag{A.8}$$

Rearrangement in terms of incremental displacements yields

$$\left[\frac{4}{h^2}\mathbf{M} + \frac{2}{h}\mathbf{C} + \mathbf{K}\right]\Delta\mathbf{x} = \mathbf{F}_{k+1} + \mathbf{M}\ddot{\mathbf{x}}_k + \left[\mathbf{C} + \frac{4}{h}\mathbf{M}\right]\dot{\mathbf{x}}_k - \mathbf{K}\mathbf{x}_k \tag{A.9}$$

This equation can then be solved for incremental displacements at each time step using the LU scheme in A.3.1. Velocities are then obtained easily by equation A.6 and to maintain equilibrium at each step, acceleration is calculated directly by substitution of these values into the equation of motion (Eq A.1) at time step $k+1$.

Stability

The stability can be investigated by considering the case of undamped free vibration of a single degree of freedom oscillator.

$$\ddot{x} + \omega^2 x = 0 \tag{A.10}$$

Rearranging equation A.9 with $F = 0$, $C = 0$, $\ddot{x}_k = -\omega^2 x_k$ and $\omega^2 = k/m$

$$\left(\frac{4}{h^2} + \omega^2\right)\Delta x = -2\omega^2 x_k + \frac{4}{h}\dot{x}_k \tag{A.11}$$

$$\Delta x = \frac{4h}{4 + h^2\omega^2}\dot{x}_k - \frac{2h^2\omega^2}{4 + h^2\omega^2}x_k \tag{A.12}$$

The displacement and velocity at $k + 1$ is then

$$
\begin{aligned}
x_{k+1} &= x_k + \Delta x \\
&= \frac{4h}{4 + h^2\omega^2}\dot{x}_k + \frac{4 - h^2\omega^2}{4 + h^2\omega^2}x_k
\end{aligned} \tag{A.13}
$$

$$
\begin{aligned}
\dot{x}_{k+1} &= \dot{x}_k + \Delta\dot{x} \tag{A.14}\\
&= \frac{2\Delta x}{h} - \dot{x}_k \\
&= \frac{4 - h^2\omega^2}{4 + h^2\omega^2}\dot{x}_k - \frac{4h\omega^2}{4 + h^2\omega^2}x_k
\end{aligned}
$$

or in difference form $Y_{k+1} = AY_k$

$$
\begin{pmatrix} \dot{x} \\ x \end{pmatrix}_{k+1} =
\begin{bmatrix}
\dfrac{4 - h^2\omega^2}{4 + h^2\omega^2} & -\dfrac{4h\omega^2}{4 + h^2\omega^2} \\
\dfrac{4h}{4 + h^2\omega^2} & \dfrac{4 - h^2\omega^2}{4 + h^2\omega^2}
\end{bmatrix}
\begin{pmatrix} \dot{x} \\ x \end{pmatrix}_k \tag{A.15}
$$

The eigenvalues of A are then calculated by $|A - \lambda I| = 0$ and simplified as

$$\lambda = \frac{4 - h^2\omega^2}{4 + h^2\omega^2} \pm \frac{4\sqrt{-h^2\omega^2}}{4 + h^2\omega^2} \tag{A.16}$$

λ is always complex as $h^2\omega^2 > 0$ and so the spectral radius is

$$
\begin{aligned}
|\lambda| &= \sqrt{\left(\frac{4 - h^2\omega^2}{4 + h^2\omega^2}\right)^2 + \left(\frac{4\sqrt{h^2\omega^2}}{4 + h^2\omega^2}\right)^2} \\
&= 1 \tag{A.17}
\end{aligned}
$$

Thus the method is *unconditionally stable*.

A.3.1 LU Factorisation Scheme

Due to the banded nature of the matrices for the shear frame structures considered, normal factorisation schemes are inefficient due to the large number of 'zero' computations. The following LU factorisation, forward substitution and backward substitution algorithms are developed to solve for incremental displacements of equation A.9. Equation A.9 is in the form $A\Delta x = b$ where

$$A = \left[\frac{4}{h^2}M + \frac{2}{h}C + K\right] \quad \text{and} \quad b = F_{k+1} + M\ddot{x}_k + \left[C + \frac{4}{h}M\right]\dot{x}_k - Kx_k \quad (A.18)$$

It is noted that A is constant for every time step whereas b varies at each step. A is therefore factorised into a lower triangular matrix L and an upper triangular matrix U. Due to the symmetric and banded nature of A, L consists of a diagonal with values all 1 and a single lower band whereas U consists of a diagonal and upper band. Thus A, L and U can be stored as shown below, where the subscripts used denote the position in the simplified system rather than the original matrices.

$$A = \begin{bmatrix} a_{1,1} & a_{1,2} & & & & 0 \\ a_{1,2} & a_{2,1} & a_{2,2} & & & \\ & \ddots & \ddots & \ddots & & \\ & & a_{n-2,2} & a_{n-1,1} & a_{n-1,2} \\ 0 & & & a_{n-1,2} & a_{n,1} \end{bmatrix} \Rightarrow A = \begin{bmatrix} a_{1,1} & a_{1,2} \\ a_{2,1} & a_{2,2} \\ \vdots & \vdots \\ a_{n-1,1} & a_{n-1,2} \\ a_{n,1} & 0 \end{bmatrix} \quad (A.19)$$

$$L = \begin{bmatrix} 1 & & & & 0 \\ l_2 & 1 & & & \\ & l_3 & 1 & & \\ & & \ddots & \ddots & \\ 0 & & & l_n & 1 \end{bmatrix} \Rightarrow L = \begin{bmatrix} l_2 \\ l_3 \\ \vdots \\ l_n \end{bmatrix} \quad (A.20)$$

$$U = \begin{bmatrix} u_{1,1} & u_{1,2} & & & 0 \\ & u_{2,1} & u_{2,2} & & \\ & & u_{3,1} & \ddots & \\ & & & \ddots & u_{n-1,2} \\ 0 & & & & u_{n,1} \end{bmatrix} \Rightarrow U = \begin{bmatrix} u_{1,1} & u_{1,2} \\ u_{2,1} & u_{2,2} \\ \vdots & \vdots \\ u_{n-1,1} & u_{n-1,2} \\ u_{n,1} & 0 \end{bmatrix} \quad (A.21)$$

The coefficients of L and U are obtained from A as

$$\text{Set } U = A \quad (A.22)$$

$$\text{for } k = 2, n$$

$$l_k = \frac{u_{k-1,1}}{u_{k-1,2}}$$

$$u_{k,1} = u_{k,1} - l_k u_{k-1,2}$$

This factorisation only needs to be carried out once as the matrices do not vary with time. At each time step, the vector **b** is calculated and the incremental displacement solved by forward and backwards substitution as follows.

Forward substitution $\mathbf{L}\mathbf{y} = \mathbf{b} \Rightarrow$ solve for **y**

$$y_1 = b_1 \tag{A.23}$$

$$for\ k = 2, n$$

$$y_k = b_k - y_{k-1}l_k$$

Backwards substitution $\mathbf{U}\mathbf{\Delta}x = \mathbf{y} \rightarrow$ solve for **Δx**

$$\Delta x_n = \frac{y_n}{u_{n,1}} \tag{A.24}$$

$$for\ k = n - 1, 1, -1$$

$$\Delta x_k = \frac{y_k - \Delta x_{k+1} u_{k,2}}{u_{k,1}}$$

Subject Index

Structures and Infrastructures Series

Book Series Editor: Dan M. Frangopol

ISSN:1747–7735

Publisher: CRC/Balkema,   Taylor & Francis Group

1. Structural Design Optimization Considering Uncertainties
   Editors: Yiannis Tsompanakis, Nikos D. Lagaros & Manolis Papadrakakis
   2008
   ISBN:978-0-415-45260-1 (Hb)

2. Computational Structural Dynamics and Earthquake Engineering
   Editors: Manolis Papadrakakis, Dimos C. Charmpis,
   Nikos D. Lagaros & Yiannis Tsompanakis
   2008
   ISBN: 978-0-415-45261-8 (Hb)

3. Computational Analysis of Randomness in Structural Mechanics
   Christian Bucher
   2009
   ISBN: 978-0-415-40354-2 (Hb)

4. Frontier Technologies for Infrastructures Engineering
   Editors: Shi-Shuenn Chen & Alfredo H-S. Ang
   2009
   ISBN: 978-0-415-49875-3 (Hb)

5. Damage Models and Algorithms for Assessment of Structures
   under Operating Conditions
   Siu-Seong Law and Xin-Qun Zhu
   ISBN: 978-0-415-42195-9 (Hb)

6. Structural Identification and Damage Detection using Genetic Algorithms
   Chan Ghee Koh and Michael John Perry
   ISBN: 978-0-415-46102-3 (Hb)